New and Emerging Proteomic Techniques

METHODS IN MOLECULAR BIOLOGY™

John M. Walker, SERIES EDITOR

352. Protein Engineering Protocols, edited by *Kristian Müller and Katja Arndt, 2006*

351. *C. elegans:* *Methods and Applications,* edited by *Kevin Strange, 2006*

350. Protein Folding Protocols, edited by *Yawen Bai and Ruth Nussinov 2006*

349. YAC Protocols, *Second Edition,* edited by *Alasdair MacKenzie, 2006*

348. Nuclear Transfer Protocols: *Cell Reprogramming and Transgenesis,* edited by *Paul J. Verma and Alan Trounson, 2006*

347. Glycobiology Protocols, edited by *Inka Brockhausen-Schutzbach, 2006*

346. *Dictyostelium discoideum* Protocols, edited by *Ludwig Eichinger and Francisco Rivero-Crespo, 2006*

345. Diagnostic Bacteriology Protocols, *Second Edition,* edited by *Louise O'Connor, 2006*

344. Agrobacterium Protocols, Second Edition: *Volume 2,* edited by *Kan Wang, 2006*

343. Agrobacterium Protocols, Second Edition: *Volume 1,* edited by *Kan Wang, 2006*

342. MicroRNA Protocols, edited by *Shao-Yao Ying, 2006*

341. Cell–Cell Interactions: *Methods and Protocols,* edited by *Sean P. Colgan, 2006*

340. Protein Design: *Methods and Applications,* edited by *Raphael Guerois and Manuela López de la Paz, 2006*

339. Microchip Capillary Electrophoresis: *Methods and Protocols,* edited by *Charles Henry, 2006*

338. Gene Mapping, Discovery, and Expression: *Methods and Protocols,* edited by *M. Bina, 2006*

337. Ion Channels: *Methods and Protocols,* edited by *James D. Stockand and Mark S. Shapiro, 2006*

336. Clinical Applications of PCR: *Second Edition,* edited by *Y. M. Dennis Lo, Rossa W. K. Chiu, and K. C. Allen Chan, 2006*

335. Fluorescent Energy Transfer Nucleic Acid Probes: *Designs and Protocols,* edited by *Vladimir V. Didenko, 2006*

334. PRINS and *In Situ* PCR Protocols: *Second Edition,* edited by *Franck Pellestor, 2006*

333. Transplantation Immunology: *Methods and Protocols,* edited by *Philip Hornick and Marlene Rose, 2006*

332. Transmembrane Signaling Protocols: *Second Edition,* edited by *Hydar Ali and Bodduluri Haribabu, 2006*

331. Human Embryonic Stem Cell Protocols, edited by *Kursad Turksen, 2006*

330. Embryonic Stem Cell Protocols, Second Edition, *Vol. II: Differentiation Models,* edited by *Kursad Turksen, 2006*

329. Embryonic Stem Cell Protocols, Second Edition, *Vol. I: Isolation and Characterization,* edited by *Kursad Turksen, 2006*

328. New and Emerging Proteomic Techniques, edited by *Dobrin Nedelkov and Randall W. Nelson, 2006*

327. Epidermal Growth Factor: *Methods and Protocols,* edited by *Tarun B. Patel and Paul J. Bertics, 2006*

326. *In Situ* Hybridization Protocols, *Third Edition,* edited by *Ian A. Darby and Tim D. Hewitson, 2006*

325. Nuclear Reprogramming: *Methods and Protocols,* edited by *Steve Pells, 2006*

324. Hormone Assays in Biological Fluids, edited by *Michael J. Wheeler and J. S. Morley Hutchinson, 2006*

323. *Arabidopsis* Protocols, *Second Edition,* edited by *Julio Salinas and Jose J. Sanchez-Serrano, 2006*

322. *Xenopus* Protocols: *Cell Biology and Signal Transduction,* edited by *X. Johné Liu, 2006*

321. Microfluidic Techniques: *Reviews and Protocols,* edited by *Shelley D. Minteer, 2006*

320. Cytochrome P450 Protocols, *Second Edition,* edited by *Ian R. Phillips and Elizabeth A. Shephard, 2006*

319. Cell Imaging Techniques, *Methods and Protocols,* edited by *Douglas J. Taatjes and Brooke T. Mossman, 2006*

318. Plant Cell Culture Protocols, *Second Edition,* edited by *Victor M. Loyola-Vargas and Felipe Vázquez-Flota, 2005*

317. Differential Display Methods and Protocols, *Second Edition,* edited by *Peng Liang, Jonathan Meade, and Arthur B. Pardee, 2005*

316. Bioinformatics and Drug Discovery, edited by *Richard S. Larson, 2005*

315. Mast Cells: *Methods and Protocols,* edited by *Guha Krishnaswamy and David S. Chi, 2005*

314. DNA Repair Protocols: *Mammalian Systems, Second Edition,* edited by *Daryl S. Henderson, 2006*

313. Yeast Protocols: *Second Edition,* edited by *Wei Xiao, 2005*

312. Calcium Signaling Protocols: *Second Edition,* edited by *David G. Lambert, 2005*

311. Pharmacogenomics: *Methods and Protocols,* edited by *Federico Innocenti, 2005*

310. Chemical Genomics: *Reviews and Protocols,* edited by *Edward D. Zanders, 2005*

309. RNA Silencing: *Methods and Protocols,* edited by *Gordon Carmichael, 2005*

308. Therapeutic Proteins: *Methods and Protocols,* edited by *C. Mark Smales and David C. James, 2005*

METHODS IN MOLECULAR BIOLOGY™

New and Emerging Proteomic Techniques

Edited by

Dobrin Nedelkov

and

Randall W. Nelson

Intrinsic Bioprobes Inc.
Tempe, AZ

HUMANA PRESS TOTOWA, NEW JERSEY

999 Riverview Drive, Suite 208
Totowa, New Jersey 07512

www.humanapress.com

This publication is printed on acid-free paper. ∞
ANSI Z39.48-1984 (American Standards Institute)

Permanence of Paper for Printed Library Materials.

Production Editor: Tracy Catanese

Cover design by Patricia F. Cleary

Cover Illustration: Fig. 7 from Chapter 7, "Surface Plasmon Resonance Imaging Measurements of Protein Interactions With Biopolymer Microarrays," by Terry T. Goodrich, Alastair W. Wark, Robert M. Corn, and Hye Jin Lee, and Fig. 9 from Chapter 3, "Antibody Microarrays Using Resonance Light-Scattering Particles for Detection," by Bernhard H. Geierstanger, Petri Saviranta, and Achim Brinker.

For additional copies, pricing for bulk purchases, and/or information about other Humana titles, contact Humana at the above address or at any of the following numbers: Tel.: 973-256-1699; Fax: 973-256-8341; E-mail: orders@humanapr.com; or visit our Website: www.humanapress.com

Printed in the United States of America. 10 9 8 7 6 5 4 3 2 1

eISBN 1-59745-026-X

Library of Congress Cataloging in Publication Data

New and emerging proteomic techniques / edited by Dobrin Nedelkov and Randall W. Nelson.
p. ; cm. -- (Methods in molecular biology ; 328)
Includes bibliographical references and index.
ISBN 1-58829-519-2 (alk. paper)
1. Proteomics--Laboratory manuals.
[DNLM: 1. Proteomics--methods--Laboratory Manuals. 2. Chemistry, Analytical--methods--Laboratory Manuals. 3. Computational Biology--methods--Laboratory Manuals. 4. Genetic Techniques--Laboratory Manuals. QU 25 N5314 2006] I. Nedelkov, Dobrin. II. Nelson, Randall W. III. Series: Methods in molecular biology (Clifton, N.J.) ; v. 328.
QP551.N49 2006
572'.6'078--dc22

2005016815

Preface

Ever since its first definition, proteomics has been referred to as the study of the full set of proteins encoded by a genome, and qualitative and quantitative comparisons of proteomes under different conditions to further unravel biological processes. Initially, proteomics was mainly a technology-driven field, centered on two-dimensional gels/mass spectrometry and LC-MS/MS approaches, which offered the ability to detect hundreds of proteins in a single analysis. Proteomics has since evolved into a wide-ranging discipline that includes a plethora of technologies and approaches whose subject is the study of proteins. However, as proteomics is entering into the realms of clinical and diagnostic applications, the enabling techniques are coming under an increased level of scrutiny. Significant weaknesses in some technological aspects have thus become apparent. As a result, new and improved proteomics techniques are being developed and propagated.

It is the intent of *New and Emerging Proteomic Techniques* to present some of the newer and still developing proteomics tools and techniques that enable enhanced protein analyses. The techniques span the entire spectrum of top-down and bottom-up approaches, and in their sum offer a clear example of how proteomics has embraced essentially all techniques that contend with protein analysis. From microarrays and gels, to chromatography and affinity separations, the proteomics techniques described in this book are addressing every aspect of the human proteome, both quantitative and qualitative. The methods of protein detection utilized are also very diverse, ranging from fluorescence and resonance light scattering, to surface plasmon resonance and mass spectrometry. Furthermore, several chapters describe a combination of two or more distinct techniques, resulting in enabling approaches for proteome analysis. There are also three chapters that describe advanced bioinformatics approaches, as they are becoming increasingly important in the analysis of the complex proteomics data.

New and Emerging Proteomic Techniques is aimed at both beginners and more experienced practitioners in the field of proteomics. Beginners will find it very useful to have such a diverse set of techniques assembled in a single book, serving as a valuable reference when choosing a technique that can address a specific proteomics question. For experienced proteomics researchers, the book offers protocols and know-how from the pioneers and the expert users of each of these techniques, with details that are usually not found in a typical research publication. We are well aware that there are other proteomics approaches that are not represented in this book. However, we feel that the fifteen chapters included describe some of the most promising new and emerging proteomics techniques, and hope that at least some of them will become proteomics mainstays in the years to come.

Finally, we would like to thank all of our colleagues who kindly contributed their time and expertise for the assembly of *New and Emerging Proteomic Techniques*.

Dobrin Nedelkov
Randall W. Nelson

Contents

Contributors

RUEDI AEBERSOLD • *Institute for Systems Biology, Seattle, WA, and Swiss Federal Institute of Technology (ETH) and Faculty of Natural Sciences, University of Zurich, Zurich, Switzerland*
RONALD C. BEAVIS • *Beavis Informatics Ltd., Winnipeg, Canada*
CELIA R. BERKERS • *Division of Cellular Biochemistry, Netherlands Cancer Institute, Amsterdam, the Netherlands*
ACHIM BRINKER • *Genomics Institute of the Novartis Research Foundation, San Diego, CA*
CHRISTOPHER M. COLANGELO • *W. M. Keck Foundation Biotechnology Resource Laboratory, Yale University, New Haven, CT*
ROBERT M. CORN • *Department of Chemistry, University of California, Irvine, CA*
GOKHAN DEMIRKAN • *Harvard Institute of Proteomics, Department of Biological Chemistry and Molecular Pharmacology, Harvard Medical School, Cambridge, MA*
MARTIN ETHIER • *Department of Chemistry, University of Manitoba, Winnipeg, Canada, and Ottawa Institute of Systems Biology, University of Ottawa, Ottawa, Canada*
DANIEL FIGEYS • *Ottawa Institute of Systems Biology, University of Ottawa, Ottawa, Canada*
BOGDAN I. FLOREA • *Leiden Institute of Chemistry, Leiden University, Leiden, the Netherlands*
LAURENCE FLORENS • *Stowers Institute for Medical Research, Kansas City, MO*
EWA FOLTA-STOGNIEW • *W. M. Keck Foundation Biotechnology Resource Laboratory, Yale University, New Haven, CT*
BERNHARD H. GEIERSTANGER • *Genomics Institute of the Novartis Research Foundation, San Diego, CA*
TERRY T. GOODRICH • *Department of Chemistry, University of California, Irvine, CA*
BRIAN B. HAAB • *The Van Andel Research Institute, Grand Rapids, MI*
EUGENIE HAINSWORTH • *Harvard Institute of Proteomics and Technology and Engineering Center, Department of Biological Chemistry and Molecular Pharmacology, Harvard Medical School, Cambridge, MA*
URBAN A. KIERNAN • *Intrinsic Bioprobes Inc., Tempe, AZ*

JOSHUA LABAER • *Department of Biological Chemistry and Molecular Pharmacology, Harvard Institute of Proteomics, Harvard Medical School, Cambridge, MA*

HYE JIN LEE • *Department of Chemistry, University of California, Irvine, CA*

XIAOYE LI • *Department of Applied Mathematics, Yale University, New Haven, CT*

JUNFENG LIU • *Department of Statistics, West Virginia University, Morgantown, WV*

PAUL M. LIZARDI • *Department of Pathology, School of Medicine, Yale University, New Haven, CT*

DOBRIN NEDELKOV • *Intrinsic Bioprobes Inc., Tempe, AZ*

RANDALL W. NELSON • *Intrinsic Bioprobes Inc., Tempe, AZ*

ERIC E. NIEDERKOFLER • *Intrinsic Bioprobes Inc., Tempe, AZ*

HUIB OVAA • *Division of Cellular Biochemistry, Netherlands Cancer Institute, Amsterdam, the Netherlands*

HERMAN S. OVERKLEEFT • *Leiden Institute of Chemistry, Leiden University, Leiden, the Netherlands*

HÉLÈNE PERREAULT • *Department of Chemistry, University of Manitoba, Winnipeg, Canada*

NIROSHAN RAMACHANDRAN • *Department of Biological Chemistry and Molecular Pharmacology, Harvard Institute of Proteomics, Harvard Medical School, Cambridge, MA*

PETRI SAVIRANTA • *Genomics Institute of the Novartis Research Foundation, San Diego, CA*

KATHY STONE • *W. M. Keck Foundation Biotechnology Resource Laboratory, Yale University, New Haven, CT*

KEMMONS A. TUBBS • *Intrinsic Bioprobes Inc., Tempe, AZ*

PAUL F. VAN SWIETEN • *Leiden Institute of Chemistry, Leiden University, Leiden, the Netherlands*

MARTIJN VERDOES • *Leiden Institute of Chemistry, Leiden University, Leiden, the Netherlands*

ALASTAIR W. WARK • *Department of Chemistry, University of California, Irvine, CA*

MICHAEL P. WASHBURN • *Stowers Institute for Medical Research, Kansas City, MO*

KENNETH R. WILLIAMS • *W. M. Keck Foundation Biotechnology Resource Laboratory, Yale University, New Haven, CT*

BAOLIN WU • *Division of Biostatistics, School of Public Health, University of Minnesota, Minneapolis, MN*

TERENCE L. WU • *W. M. Keck Foundation Biotechnology Resource Laboratory, Yale University, New Haven, CT*

WEICHUAN YU • *Department of Molecular Biophysics and Biochemistry, Yale University, New Haven, CT*
HUI ZHANG • *Institute for Systems Biology, Seattle, WA*
HONGYU ZHAO • *Department of Epidemiology and Public Health, Yale University, New Haven, CT*

1

On-Chip Protein Synthesis for Making Microarrays

Niroshan Ramachandran, Eugenie Hainsworth, Gokhan Demirkan, and Joshua LaBaer

Summary

Protein microarrays are a miniaturized format for displaying in close spatial density hundreds or thousands of purified proteins that provide a powerful platform for the high-throughput assay of protein function. The traditional method of producing them requires the high-throughput production and printing of proteins, a laborious method that raises concerns about the stability of the proteins and the shelf life of the arrays. A novel method of producing protein microarrays, called *nucleic acid programmable protein array* (NAPPA), overcomes these limitations by synthesizing proteins *in situ*. NAPPA entails spotting plasmid DNA encoding the relevant proteins, which are then simultaneously transcribed and translated by a cell-free system. The expressed proteins are captured and oriented at the site of expression by a capture reagent that targets a fusion protein on either the N- or C-terminus of the protein. Using a mammalian extract, NAPPA expresses and captures 1000-fold more protein per feature than conventional protein-printing arrays. Moreover, this approach minimizes concerns about protein stability and integrity, because proteins are produced just in time for assaying. NAPPA has already proven to be a robust tool for protein functional assays.

Key Words: Protein microarrays; functional proteomics; protein expression; protein purification; microarray surface chemistry.

1. Introduction

The recent development of functional protein microarrays has stirred excitement in the proteomics community ***(1–7)***. The power of this approach is that, by spotting many proteins on a single array surface, many biochemical activities can be studied simultaneously. These activities include identifying interacting proteins, examining the selectivity of drug binding, finding substrates for active enzymes, and looking for unintended drug interactions. Typically, the array is probed with a labeled query molecule to identify interactions with proteins on

From: *Methods in Molecular Biology, Vol. 328: New and Emerging Proteomic Techniques*
Edited by: D. Nedelkov and R. W. Nelson © Humana Press Inc., Totowa, NJ

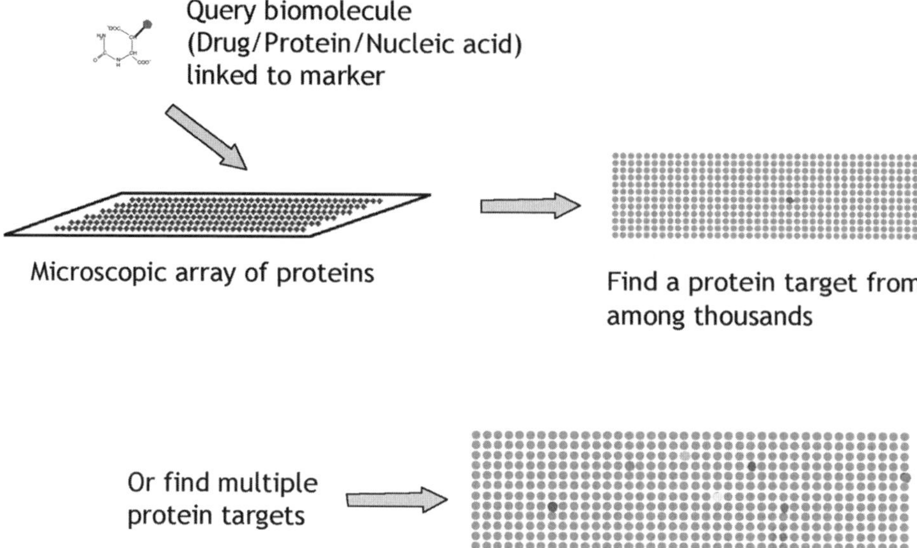

Fig. 1: Rapid screening of target protein arrays. A labeled query molecule can be used to screen an array of target proteins to identify potential interactor(s).

the array (**Fig. 1**). For example, a labeled candidate kinase inhibitor might be used to screen an array of kinases to determine which kinase(s) the inhibitor binds directly. In order to build protein microarrays, one needs the content to spot on the array and an appropriate binding chemistry to capture the protein. These components must be optimized to produce and present proteins of good integrity and stability. The goal is to preserve the functionality of the protein in order to minimize false-negatives. Here we will address the issues pertaining to building functional protein microarrays.

1.1. Issues for Protein Array Production

1. *Availability of array content.* Assembling the proteins for printing on the array remains a major challenge for most researchers, because recombinant expression of proteins in the numbers anticipated for protein microarrays relies on the availability of large collections of cDNAs in protein expression-ready formats. It also requires methods to produce and purify the proteins. Although several collections of cDNAs are available, the methods and robotic equipment required for the high-throughput (HT) expression and purification of thousands of proteins remain outside the realm of most laboratories ***(8–10)***.
2. *Protein integrity.* Ensuring that properly folded proteins can be produced and captured remains a challenge. Proteins are more likely to fold naturally if heterologous

systems can be avoided and if proteins are synthesized in a milieu as close to their natural setting as possible. For example, mammalian proteins are more likely to fold naturally in mammalian (or at least animal) cells. Yet, many expression systems are too cumbersome and expensive to allow thousands of proteins to be easily processed. Bacterial cells, which are readily adapted to HT protein expression, can be counted on to produce only 50% of mammalian proteins ***(10)***.

3. *Protein stability.* Proteins are notoriously fragile, raising concerns about the stability of the isolated proteins before and after they have been arrayed on a glass slide. We expect some proteins to remain relatively stable, with good shelf life, whereas others display greater lability and are unable to withstand prolonged array conditions. Moreover, it is difficult to determine which of the proteins remain active at the time of the assay. In general, to ensure proper assay conditions and minimize false-negatives, it is best to use the array soon after the protein is synthesized.
4. *Microarray surface chemistry.* Several surface chemistries have been developed and validated for microarray platforms, particularly for DNA microarrays. The chemistry for binding DNA is simple compared with the chemical demands necessary to immobilize functional protein. DNA molecules, which are all negatively charged, bind to surfaces based on charge alone, enabling positively charged arrays such as a polylysine-coated slides to bind all DNA. In contrast, proteins display a staggering range of hydrophobicity and charge, making it a challenge to find a single method that provides good binding for most, let alone all, proteins. Factors to consider include:
   a. Generality of binding: ability to bind all proteins that will be spotted on the array.
   b. Binding capacity: maximum amount of protein captured per feature.
   c. Efficiency of capture: fraction of spotted protein that is captured on the array.
   d. Orientation: specific vs random orientation. Proteins can be immobilized either in an orientation-specific manner (e.g., by binding via either an N-terminus or a C-terminus tag) or in random orientations (e.g., by chemical attachment). Random rather than specific orientations may allow many areas of the protein to be exposed, increasing accessibility to the protein. Although this may increase the likelihood of an interaction, there have been no significant differences observed between these approaches ***(11)***; it may be necessary to evaluate this on an experiment by experiment basis.
   e. Distance from surface: some attachment methods allow for a spacer (e.g., a large polypeptide tag) that separates the protein from the array surface; other methods (e.g., chemical attachment) bring the proteins in direct contact with the array surface. Increasing the distance between the protein and the array surface might alleviate some of the steric hindrance caused by the surface and potentially increase accessibility to the protein.
   f. Native or denatured protein: surface chemistry can be formulated to contain hydrophobic or hydrophilic residues. Given that many proteins have a hydrophilic exterior and a hydrophobic interior, the choice of the surface chemistry could support the binding of nondenatured or denatured protein ***(12)***.

Early demonstration of the feasibility of printing proteins on a microscopic surface has been promising. To demonstrate that spotted proteins maintain their functional integrity upon immobilization, well characterized and specific interactions among proteins, lipids, and small molecules, as well as enzyme-substrate screens, were recapitulated with proteins on the arrays ***(1,3,6,7,11,13–20)***. Even in light of these achievements, the widespread use of this technology has remained limited, largely as a result of the labor-intensive protein production, the quality of proteins expressed in heterologous systems, and the stability of the proteins during storage. To address these persistent concerns, we developed a self-assembling protein microarray method.

1.2. Nucleic Acid Programmable Protein Array

To circumvent the need to express, purify, and spot the protein, this approach prints the plasmids bearing the genes on the array and the proteins are synthesized *in situ*. The genes are configured such that each expressed protein contains a polypeptide tag used to capture the protein to the array surface. The proteins are expressed using a cell-free transcription/translation extract, which can be selected to match the source of the genes (e.g., rabbit reticulocyte lysate for mammalian genes), thus enabling the proteins to be expressed in a more native milieu. The use of appropriate cell-free extracts helps to encourage natural folding and, at least in the case of reticulocyte lysate, is highly successful at expressing most proteins. In addition, some natural posttranslational modifications occur in these extracts and/or can be induced by using supplemented lysates ***(21,22)***.

Arranging the genes so that each has an appropriate capture tag is facilitated by using vectors with recombinational cloning sites. Coding regions inserted in recombinational cloning systems, such as the Invitrogen Gateway system or Clontech Creator system, can be readily moved into expression vectors that append the appropriate tag(s) to the coding regions. The transfer reactions themselves are simple, highly efficient, error-free, and automatable. The assembly of large collections of genes in these systems is currently in progress ***(10,23–27)***.

A significant advantage of the nucleic acid programmable protein array (NAPPA) approach is that it eliminates concerns about protein stability. Proteins on the array are not produced until the array is ready for use in experiments; that is, they are made just in time. Prior to activation with the cell-free transcription/translation extract, the arrays are stable and can be stored dry on the bench for months.

Using this approach in a recent study, 30 human DNA replication proteins were expressed and captured on NAPPA microarrays ***(28)***. The yield of captured protein was 400–2700 pg/feature, which was 1000-fold more than conventional protein-spotting arrays, 10–950 fg/feature ***(11)***. Arrays were used to determine protein-protein interactions (recapitulating 85% of the previously known interactions), to map protein interaction domains by using partial-length proteins, and to assemble multiprotein complexes.

2. Materials

2.1. Equipment

1. Arrayer with solid pins, humidity control.
2. Microarray scanner.
3. Programmable chilling incubator.
4. SpeedVac.
5. Centrifuge: Sorvall RC12, Eppendorf 5417C, IEC Centra GP8.
6. Ultraviolet (UV) light, UVP UVLMS-38, set at 365 nm.

2.2. Preparation of the Slides

1. Glass slides (VWR 48311-702).
2. Solution of 2% aminosilane (Pierce 80370) in acetone. Make up 300 mL just before use.
3. Stainless steel 30-slide rack (Wheaton), handle removed.
4. Glass staining box (Wheaton).
5. Lock & Lock 1.5 cup boxes (Heritage Mint Ltd., ZHPL810).
6. Prepare a 50 m*M* dimethyl suberimidate·2 HCl (DMS) stock solution: 1 g of DMS linker (Pierce 20700) in 40 mL dimethylsulfoxide (DMSO). Store at –20°C.
7. To coat slides with linker only (used if NAPPA strategy is to spot avidin/streptavidin along with plasmid DNA and anti-glutathione *S*-transferase [GST] antibody): 2 m*M* DMS in phosphate-buffered saline (PBS), pH 9.5 (*see* **Note 1**).

   Or
8. To coat slides with avidin/streptavidin (used if NAPPA strategy is to spot only plasmid DNA and anti-GST antibody): 2 m*M* DMS, plus avidin (Cortex CE0101) at 1 mg/mL or streptavidin (Cortex CE0301) at 3.5 mg/mL, in PBS, pH 9.5. For material in either **step 7** or **8**, make fresh at the time of coating, otherwise the DMS linker may hydrolyze over time (*see* **Note 1**).
9. Cover slips (VWR 48393–081).
10. Bioassay dishes with dividers (Genetix x6027).

2.3. DNA Preparation

1. The plasmid DNA is prepared in 300-mL cultures usually grown in Terrific Broth media. The DNA preparation is derived from Sambrook et al. ***(29)*** and is summarized below.
2. Solution 1 (GTE): 50 m*M* glucose, 25 m*M* Tris-HCl (pH 8.0), 10 m*M* ethylenediamine tetraacetic acid (EDTA), pH 8.0, and 0.1 mg/mL RNAse. Store at 4°C.
3. Solution 2: 0.2 *N* NaOH with 1% sodium dodecyl sulfate (SDS).
4. Solution 3: 3 *M* KOAC; add glacial acetic acid until pH is 5.5

5. 250-mL conical Corning centrifuge bottle.
6. Glass fiber 0.7-µm filter plate, long drip (Innovative Microplate F20060).
7. 96-well deep-well block (Marsh AB-0661).

2.4. Preparation of Samples and Arraying

1. Plasmid DNA (prepared in **Subheading 2.3.**).
2. Microcon YM-100 (100 kDa) tube (Millipore), or DNA binding plate: 100 kDa 96-well filter plate (Millipore plasmid plate).
3. BrightStar Psoralen-biotin kit (Ambion 1480). Just before use, prepare psoralen-biotin: dissolve the contents (4.17 ng) of the kit in 50 µL DMF (also in kit).

Or

4. EZ-Link Psoralen-PEO-Biotin (Pierce 29986). Prepare stock solution of 5 mg/mL in water and store at –20°C.
5. UV-transparent 96-well plate (Corning 3635).
6. Sephadex G50 (Sigma-Aldrich).
7. 1.2-µm glass fiber filter plate, long drip (Innovative Microplate F20021).
8. Collection plate, round bottom (Corning 3795).
9. 384-well plate for arraying (Genetix x7020).
10. Polyclonal anti-GST antibody (Amersham Biosciences 27457701).
11. Purified GST protein (Sigma G5663). Prepare stock solution of 0.03 mg/mL in PBS.
12. Whole mouse immunoglobulin (Ig)G antibody (Pierce 31204). Prepare stock solution of 0.5 mg/mL in PBS.
13. Bis(sulfosuccinimidyl) suberate (BS3) linker (Pierce 21580).
14. Bioassay dish dividers to be used as slide racks (Genetix x6027) and deeper bioassay dishes (e.g., Corning 431111 or 431272; do not use "low profile" dishes).

2.5. Expression of Proteins

1. HybriWell gaskets (Grace HBW75).
2. Cell-free expression system (Rabbit reticulocyte lysate) (Promega L4610).
3. RNaseOUT (Invitrogen 10777–019).
4. SuperBlock blocking solution in TBS (Pierce 37535).
5. Milk blocking solution: 5% milk in PBS with 0.2% Tween-20 (Sigma).

2.6. Detection and Analysis

1. Primary AB solution: mouse anti-GST (Cell Signaling 2624) 1:200 in SuperBlock (Pierce 37535). Store at 4°C.
2. Primary AB solution: mouse anti-HA (Cocalico) 1:1000 in SuperBlock. Store at 4°C.
3. Secondary AB solution: horseradish peroxidase (HRP)-conjugated anti-mouse (Amersham NA931) 1:200 in SuperBlock. Store at 4°C.
4. Tyramide Signal Amplification (TSA) stock solution: use TSA reagent (PerkinElmer SAT704B001EA). Prepare per kit directions. Keep this solution at 4°C.
5. Milk blocking solution: 5% milk in PBS with 0.2% Tween-20 (Sigma).

6. Cover slips (VWR 48393–081).
7. PicoGreen (Molecular Probes P11495) stock solution: to the 100 µL/vial that comes in kit, add 200 µL TE buffer. Before use, do a 1:600 dilution in SuperBlock.

3. Methods

NAPPA chemistry relies on efficient immobilization of plasmid DNA onto a solid surface without compromise to integrity, and on rapid capture of the expressed target proteins. In order to immobilize the plasmid, we use a psoralen-biotin bis-functional linker that derivatizes the plasmid DNA (**Fig. 2**). Under long-wave UV (365 nm), psoralen intercalates into the DNA, creating a biotinylated plasmid. The reaction is fairly robust over a wide range of pH and salt concentrations. The biotinylated plasmid is tethered to the array surface by high-affinity binding to either avidin or streptavidin. In addition to the plasmids, target protein capture molecules are also immobilized on the slide. Currently, our plasmids are programmed to express target proteins with a C-terminal GST fusion protein; therefore, a polyclonal anti-GST antibody is bound to the array as the capture molecule to immobilize the expressed target proteins (**Fig. 3**). The presence of the C-terminal fusion tag can later be confirmed by incubating the slides with an antibody that recognizes a different epitope on the tag than the antibody used for capture. The presence of the C-terminal tag indicates that the full-length protein was expressed.

In order to make this chemistry robust and reproducible, we have used high-affinity capture reagents that are well characterized and stable throughout arraying and storage. Moreover, the schemes outlined previously can be altered by the user to accommodate different immobilization chemistries for the plasmid DNA and/or target proteins.

3.1. Preparation of the Slides

1. Prepare 300 mL of aminosilane coating solution (2% aminosilane reagent in acetone).
2. Put slides in metal rack (30-slide Wheaton rack).
3. Treat glass slides in the aminosilane coating solution, approx 1–15 min in glass staining box on shaker. Rinse with acetone in rack using wash bottle. Briefly rinse with Milli-Q water. Spin dry in SpeedVac or dry using 0.2-µm filtered air cans or use house air with 2 × 0.25 µm filters (*see* **Note 2**). It is important to use clean air to dry slides in order to prevent contaminating debris from binding to the surface.
4. Store at room temperature in metal rack in Lock & Lock box.
5. Just before use, prepare linker solution according to **Subheading 2.2. step 7** or **step 8**, depending on the array strategy.
6. Set slides on divider in bioassay dish, with water in the bottom of the tray. Treat each slide with 150–200 µL linker solution and cover slip (*see* **Note 3**). Incubate for 2–4 h at room temperature or overnight in cold-room.

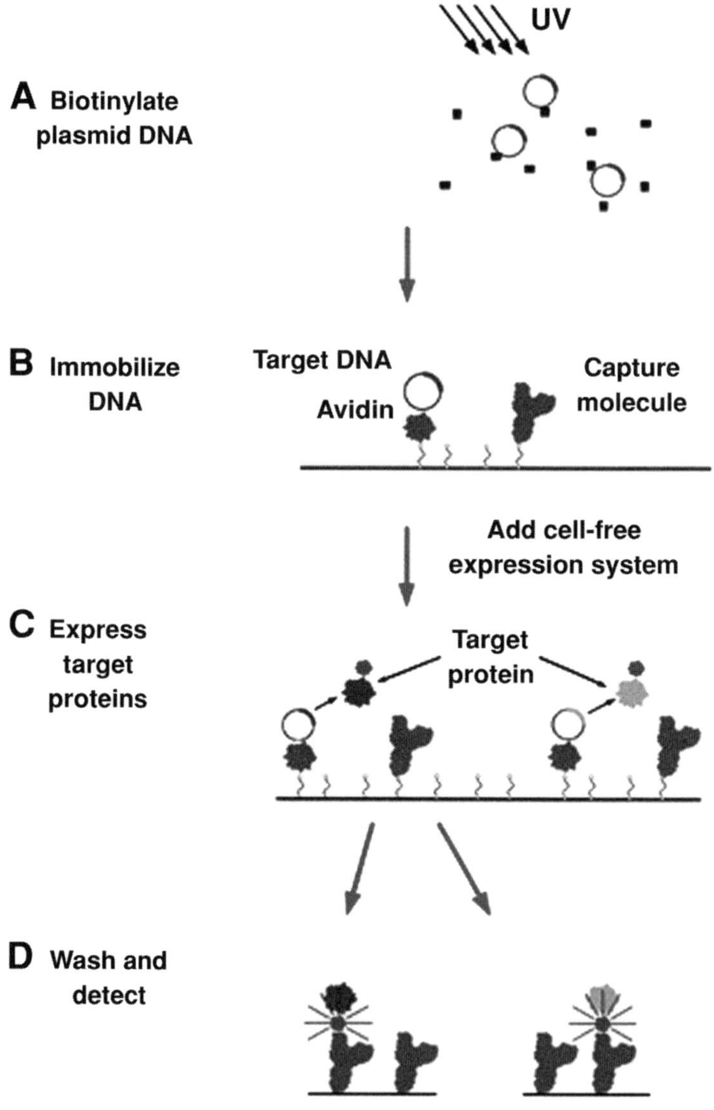

Fig. 2. Nucleic acid programmable protein array (NAPPA) chemistry. **(A)** Derivatization of plasmid DNA. Plasmid DNA is mixed with psoralen-biotin, and cross-linked using ultraviolet light. **(B)** Sample preparation. The DNA is mixed with avidin/streptavidin, crosslinker, and the anti-glutathione *S*-transferase (GST) capture antibody, and this mix is arrayed on the aminosilane-coated glass slides. **(C)** Protein expression. After blocking, cell-free expression mix is applied to the slide, and during a temperature-programmed incubation the proteins are produced and bind to the capture antibody. **(D)** Detection. The slide is washed, and the proteins are detected by detecting the GST tag (using a monoclonal anti-GST antibody, an horseradish peroxidase (HRP)-labeled anti-mouse antibody, and Cy3-tyramide [TSA] HRP substrate).

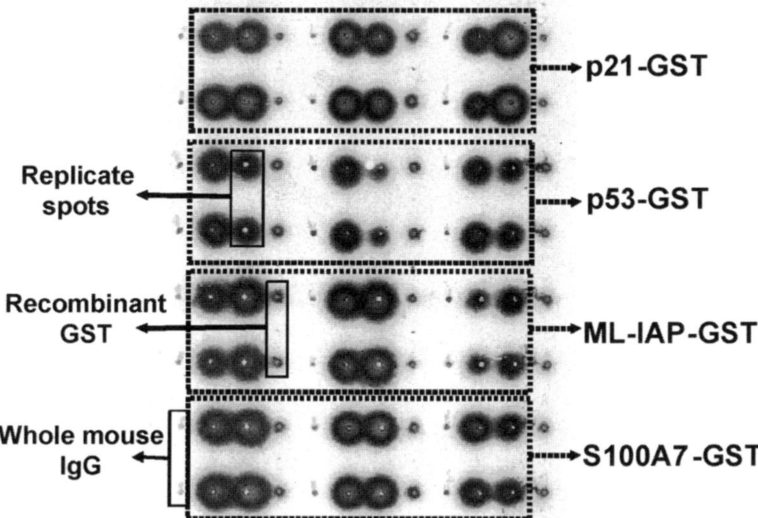

Fig. 3. Protein expression on nucleic acid programmable protein array (NAPPA) arrays. Plasmids encoding for four control proteins (p21, p53, ML-IAP, and S100A7) were biotinylated and immobilized via streptavidin onto the array surface. Arrays are activated by adding rabbit reticulocyte lysate, and the expressed proteins are detected using α-glutathione *S*-transferase (GST) antibody. Protein expression and its immediate replicate using the same pin from the same well are indicated on the left. The neighboring spots of the same gene were arrayed from different wells with different pins. Two registration spots are also included—purified recombinant GST protein and whole mouse immunoglobulin G.

7. Wash with Milli-Q water.
8. Put slides in metal rack. Spin dry in SpeedVac.
9. Store at room temperature in metal rack in Lock & Lock box.

3.2. DNA Preparation

1. Grow 300-mL culture: in a 2-L culture flask, make a 300-mL culture of TB with 10% KPI. Add 300 μL 100 mg/mL ampicillin stock solution. Add 0.5 μL glycerol stock. Put it on a shaker for 16–24 h at 37°C, 300 rpm.
2. Pellet in 450-mL centrifuge bottle: spin 15 min at 5300*g* (Sorval RC12).
3. Add 30 mL of solution 1 and resuspend.
4. Add 60 mL of solution 2 and swirl, no more than 5 min.
5. Add 45 mL of solution 3 and shake briefly.
6. Spin 15 min at 5300*g* (Sorval RC12).
7. Pass through cheesecloth into 250-mL conical Corning centrifuge bottles.
8. Add 75 mL of isopropanol and shake.
9. Spin at 5300*g* 15 min (Sorval RC12).

10. Pour off supernatant.
11. Dissolve pellet in 2 mL of Tris-EDTA buffer (pH 8.0) and transfer to a 2-mL microfuge tube. Plasmid DNA yield from this preparation is approx 0.5–1.5 μg/μL.
12. Add 200–250 μL to each well of the long drip glass fiber 0.7-μm filter plate (F20060). Stack on top of a deep-well block.
13. Spin 20 min at 890*g* (IEC Centra GP8).
14. Store the filtrate in the deep-well block at –20°C, or in individual microfuge tubes.

3.3. Preparation of Samples and Arraying

1. Either spin 200 μL of DNA (0.5–1.5 μg/μL) in a Microcon 100-kDa tube at 1000*g* for 20 min, or spin 200 μL of DNA in a 100-kDa 96-well filter plate, stacked on top of a discard plate, for 20 min at 890*g* (IEC Centra GP8).
2. Resuspend in 100 μL water. DNA concentration should be 1–2 μg/μL. The goal is to achieve 100 μL of roughly 1 μg/μL of plasmid DNA. This is because the following UV exposure conditions for biotinylation of the plasmid have been optimized for a 100-μL volume. Increasing or decreasing the volume is feasible, but the height of the liquid in the well may affect the UV dose. This may require a re-optimization of UV time and biotin dose to achieve efficient intercalation of the psoralen.
3. Just before use, prepare the BrightStar psoralen-biotin (*see* **Subheading 2.4., step 3**): dissolve the contents (4.17 ng) of the kit in 50 μL DMF (also in kit); or for EZ-Link Psoralen-PEO-Biotin (*see* **Subheading 2.4., step 4**) prepare a 0.25 mg/mL solution in water.
4. Add the resuspended DNA into a UV plate for UV cross-linking. Add 1.3 μL of BrightStar psoralen-biotin or 2 μL of 0.25 mg/mL EZ-Link Psoralen-PEO-Biotin solution per 100 μL DNA.
5. Cross-link for 20 min for BrightStar psoralen-biotin or for 30 min for EZ-Link Psoralen-PEO-Biotin with 365 nm UV, with the plate right up to the light; plate on ice; entire setup covered with foil. (The light covers five columns of the plate, so use only five columns of wells.) Note: 30 min with this setup corresponds to 8000 mJ/cm^2.
6. Prepare Sephadex slurry, 25–50 mg/mL in water. Add 200 μL of slurry to a 1.2-μm glass fiber filter plate. Spin briefly at 890*g* (IEC Centra GP8) for 1 min into a discard plate. Add 100 μL of water to the filter plate for the Sephadex to swell. Add 100 μL of DNA and spin briefly again into the collection plate. Add 100 μL water to the filter plate and spin briefly into the collection plate again.
7. Add eluate (approx 250 μL) to either a Microcon 100-kDa tube or a 100-kDa 96-well filter plate stacked on top of a discard plate. For the Microcon tube, spin at 1000*g* for 20 min (Eppendorf 5417C). For the filter plate, spin for 20 min at 890*g* (IEC Centra GP8).
8. Resuspend in 50 μL water (2 μg/μL plasmid DNA). Check that OD_{260} at 1:300 dilution is approx 0.6 (the absorbance reading is applicable only with the above mentioned method of DNA preparation; different DNA preparation methods yield different purity with different absorbance). Note: the desired final plasmid DNA concentration depends on the level of expression for the particular gene of interest.

Final plasmid DNA concentration may vary from 0.5 μg/μL for genes with good expression to 3 μg/μL for genes with poor expression.

9. Prepare spotting mix in arraying plate: 10 μL DNA + 1.5 μL of master mix.

   Master mix: *For linker-only slides*: GST polyclonal AB (0.5 mg/mL) + BS3 crosslinker (2 m*M*) + avidin (1 mg/mL) or streptavidin (3.5 mg/mL). *For avidin/streptavidin-coated slides:* GST polyclonal AB (0.5 mg/mL) + BS3 crosslinker (2 m*M*).
10. GST registration spots: 0.03 mg/mL in water or PBS.
11. Mouse IgG registration spots (whole mouse IgG antibody): 0.5 mg/mL in water or PBS.
12. Spin down plate, 1 min at 210*g* (IEC Centra GP8).
13. Array, using humidity control at 40–60%.
14. Store spotted slides in cold-room with water in the bottom of the tray, at least overnight. The bioassay dish divider should be placed in a deeper bioassay dish, so that the slides can be placed face-up on the rack without hitting the cover. Water in the bottom of the tray maintains high humidity.
15. Store slides the next day at room temperature. Storage conditions have been tested at room temperature to –80°C in the dark for up to 2 mo without loss in expression and capture.

3.4. Expression of Proteins

1. Block slides for approx 1 h at room temperature or 4°C overnight in the cold-room with SuperBlock or milk. Use approx 30 mL in a pipet box for four slides. The slides need to be shaken during this initial step to wash away unbound NAPPA reagents (plasmid, avidin/streptavidin, capture antibody).
2. Quickly rinse with milli-Q water. Dry with filtered compressed air. Do not let slides stand to dry, as the watermarks will increase background.
3. Prepare in-vitro transcription/translation (IVT) mix. For one slide, 100 μL is needed (*see* **Note 4**):
   a. 4 μL TNT buffer.
   b. 2 μL T7 polymerase.
   c. 1 μL of –Met.
   d. 1 μL of –Leu or –Cys.
   e. 2 μL of RNaseOUT.
   f. 40 μL of diethylpyrocarbonate (DEPC)-treated water.
4. Apply a HybriWell gasket to each slide. Use the wooden stick to rub the areas where the adhesive is, to make sure it is well stuck all around.
5. Add IVT mix from the nonspecimen end. Pipet the mix in slowly; it may bead up temporarily at the inlet end. Gently massage the HybriWell to get the IVT mix to spread out and cover all of the area of the array. Apply the small, round port seals to both ports.
6. Incubate for 1.5 h at 30°C for protein expression (30 is key; 28 or 32 gives reduced yield), followed by 30 min at 15°C for the query protein to bind to the immobilized protein.

7. Remove the HybriWell; wash with milk three times, 3 min each, in pipet box on a shaker. Use approx 30 mL per wash.
8. Block with SuperBlock or milk overnight at 4°C or room temperature for 1 h.

3.5. Detection and Analysis

1. Apply primary AB (mouse anti-GST or mouse anti-HA) by adding 150 µL to the nonspecimen end of the slide, then apply a cover slip. Incubate for 1 h at room temperature; wash with milk (three times, approx 5 min). Drain.
2. Apply secondary AB (anti-mouse HRP) by adding 150 µL to the nonspecimen end of the slide, then apply a coverslip. Incubate for 1 h at room temperature; wash with PBS (three times, approx 5 min). Then do a quick rinse with Milli-Q water. Drain.
3. Before applying TSA solution, make sure that the slides are not too wet, but do not let them fully dry. (If they are too wet, it will dilute the TSA.) Apply TSA mix and place cover slip. Incubate for 10 min at room temperature. Rinse in Milli-Q water; dry with filtered compressed air.
4. Scan in microarray scanner, using settings for Cy3.

As a quality check, select a couple of slides per arraying batch, and detect the arrayed DNA:

5. Block with SuperBlock 1 h.
6. For a single slide: apply 150 µL PicoGreen mix and apply cover slip. Let sit for 5 min at room temperature. For four slides, add 20 mL in a box and shake for 5 min.
7. Wash with PBS (three times, approx 5 min). Then do a quick rinse with Milli-Q water.
8. Dry with filtered compressed air.
9. Scan, using Cy3 settings.

4. Notes

1. Part of the slide preparation process involves coating the slide with an activated *N*-hydroxysuccinimide (NHS) ester cross-linker (DMS). In our experience, this has not affected protein binding of the arrayed sample, but it significantly reduces the background.
2. We have used both streptavidin and avidin to immobilize the DNA onto the array surface. We have also coated the slides with avidin or streptavidin instead of adding it to the array mixture. Avidin in the array mixture tends to precipitate and affects arraying of the sample, whereas streptavidin does not precipitate and hence is a better choice for adding to the array mixture. Coating the slides with either streptavidin or avidin is feasible; however, we do observe better binding when the proteins are in the array mixture.
3. A key step in processing microarray slides is to *never* let them air dry. Any attempt to dry them slowly will leave watermarks, which will result in high background. As suggested in the protocol, use a clean air source to quickly dry the slides. It is also important to rinse the slides in clean, filtered water before drying, especially if the

arrays have been incubating in salt or protein solutions. Drying arrays directly from salt/protein solutions can also generate high background.

4. Good expression of the NAPPA arrays depends on correctly preparing the rabbit reticulocyte lysate. It is advisable to test a small sample of your prepared lysate for expression using the positive control provided in the kit.

Acknowledgments

We thank Todd Golub's Cancer Genomics lab at the Broad Institute for allowing us to use their microarrayer and scanner. This project was funded by the National Institutes of Health/National Cancer Institute grant for functional proteomics of Breast Cancer (R21 CA99191-01).

References

1. Mitchell, P. (2002) A perspective on protein microarrays. *Nat. Biotech.* **20,** 225–229.
2. Ramachandran, N. and LaBaer, J. (2005) Protein microarrays as tools for functional proteomics. *Curr. Opin. Chem. Biol.* **9(1),** 14–19.
3. MacBeath, G. (2002) Protein microarrays and proteomics. *Nat. Genet.* **32 Suppl.,** 526–532.
4. Jona, G. and Snyder, M. (2003) Recent developments in analytical and functional protein microarrays. *Curr. Opin. Mol. Ther.* **5,** 271–277.
5. Predki, P. F. (2004) Functional protein microarrays: ripe for discovery. *Curr. Opin. Chem. Biol.* **8,** 8–13.
6. Cahill, D. J. and Nordhoff, E. (2003) Protein arrays and their role in proteomics. *Adv. Biochem. Eng. Biotechnol.* **83,** 177–187.
7. Zhu, H. and Snyder, M. (2003) Protein chip technology. *Curr. Opin. Chem. Biol.* **7,** 55–63.
8. Murthy, T. V., Wu, W., Qiu, Q. Q., Shi, Z., LaBaer, J., and Brizuela, L. (2004) Bacterial cell-free system for high-throughput protein expression and a comparative analysis of Escherichia coli cell-free and whole cell expression systems. *Protein Expr. Purif.* **36,** 217–225.
9. Holz, C., Prinz, B., Bolotina, N., Sievert, V., Bussow, K., Simon, B., et al. (2003) Establishing the yeast Saccharomyces cerevisiae as a system for expression of human proteins on a proteome-scale. *J. Struct. Funct. Genomics* **4,** 97–108.
10. Braun, P., Hu, Y., Shen, B., Halleck, A., Koundinya, M., Harlow, E., et al. (2002) Proteome-scale purification of human proteins from bacteria. *Proc. Natl. Acad. Sci. USA* **99,** 2654–2659.
11. Zhu, H., Bilgin, M., Bangham, R., Hall, D., Casamayor, A., Bertone, P., et al. (2001) Global analysis of protein activities using proteome chips. *Science* **293,** 2101–2105.
12. Mrksich, M. and Whitesides, G. M. (1996) Using self-assembled monolayers to understand the interactions of man-made surfaces with proteins and cells. *Annu. Rev. Biophys. Biomol. Struct.* **25,** 55–78.

13. MacBeath, G. and Schreiber, S. (2000) Printing proteins as microarrays for high-throughput function determination. *Science* **289,** 1760–1763.
14. Newman, J. R. and Keating, A. E. (2003) Comprehensive identification of human bZIP interactions with coiled-coil arrays. *Science* **300,** 2097–2101.
15. Haab, B., Dunham, M., and Brown, P. (2001) Protein microarrays for highly parallel detection and quantitation of specific proteins and antibodies in complex solutions. *Genome Biol.* **2(2),** RESEARCH0004.
16. Haab, B. B. (2003) Methods and applications of antibody microarrays in cancer research. *Proteomics* **3,** 2116–2122.
17. Boutell, J. M., Hart, D. J., Godber, B. L., Kozlowski, R. Z., and Blackburn, J. M. (2004) Functional protein microarrays for parallel characterisation of p53 mutants. *Proteomics* **4,** 1950–1958.
18. Prime, K. L. and Whitesides, G. M. (1991) Self-assembled organic monolayers: model systems for studying adsorption of proteins at surfaces. *Science* **252,** 1164–1167.
19. Michaud, G. A., Salcius, M., Zhou, F., Bangham, R., Bonin, J., Guo, H., et al. (2003). Analyzing antibody specificity with whole proteome microarrays. *Nat. Biotechnol.* **21,** 1509–1512.
20. Miller, J. C., Zhou, H., Kwekel, J., Cavallo, R., Burke, J., Butler, E. B., et al. (2003) Antibody microarray profiling of human prostate cancer sera: antibody screening and identification of potential biomarkers. *Proteomics* **3,** 56–63.
21. Starr, C. and Hanover, J. (1990) Glycosylation of nuclear pore protein p62. Reticulocyte lysate catalyzes O-linked N-acetylglucosamine addition in vitro. *J. Biol. Chem.* **265,** 6868–6873.
22. Walter, P. and Blobel, G. (1983) Preparation of microsomal membranes for cotranslational protein translocation. *Methods Enzymol.* **96,** 84–93.
23. Brizuela, L., Braun, P., and LaBaer, J. (2001) FLEXGene repository: from sequenced genomes to gene repositories for high-throughput functional biology and proteomics. *Mol. Biochem. Parasitol.* **118,** 155–165.
24. Walhout, A., Temple, G., Brasch, M., Hartley, J., Lorson, M., van den Heuvel, S., et al. (2000) GATEWAY recombinational cloning: application to the cloning of large numbers of open reading frames or ORFeomes. *Methods Enzymol.* **328,** 575–592.
25. Walhout, A., Sordella, R., Lu, X., Hartley, J., Temple, G., Brasch, M., et al. (2000) Protein interaction mapping in C. elegans using proteins involved in vulval development. *Science* **287,** 116–122.
26. Labaer, J., Qiu, Q., Anumanthan, A., Mar, W., Zuo, D., Murthy, T. V., et al. (2004) The *Pseudomonas aeruginosa* PA01 gene collection. *Genome Res.* **14,** 2190–2200.
27. Aguiar, J. C., LaBaer, J., Blair, P. L., Shamailova, V. Y., Koundinya, M., Russell, J. A., et al. (2004) High-throughput generation of P. falciparum functional molecules by recombinational cloning. *Genome Res.* **14,** 2076–2082.
28. Ramachandran, N., Hainsworth, E., Bhullar, B., Eisenstein, S., Rosen, B., Lau, A. Y., et al. (2004) Self-assembling protein microarrays. *Science* **305,** 86–90.
29. Sambrook, J., Fritsch, E. F., and Maniatis, T., eds. (1989) *Molecular Cloning . A laboratory manual.* Cold Spring Harbor Laboratory, Cold Spring Harbor, NY.

2

RCA-Enhanced Protein Detection Arrays

Brian B. Haab and Paul M. Lizardi

Summary

There are many instances in which it is desirable to generate profiles of the relative abundance of a multiplicity of protein species. Examples include studies in embryonic development, immunobiology, drug responses, cancer biology, biomarkers, and so on. Microarray formats provide a convenient, high-throughput vehicle for generating such profiles, and the repertoire of proteins that can be measured is growing continuously as larger panels of specific antibodies become available. Here we describe methods for the use of antibody microarrays, whereby the detection of specifically bound antigens is enhanced by rolling circle amplification (RCA). RCA-enhanced protein detection on antibody microarrays provides a means for rapid protein profiling at high sensitivity. The set of RCA reagents remains unchanged for different microarray formats and compositions, and signal readout is performed using standard fluorescent dyes and scanners. The method is sensitive enough for the most challenging applications, such as the detection of low-abundance components of human serum.

Key Words: Microarrays; proteomics; rolling circle; antibody arrays; signal amplification; protein profiling.

1. Introduction

The feasibility and utility of antibody microarrays for the highly multiplexed analysis of proteins is now well established ***(1–11)***. Two major types of antibody microarray detection systems have emerged: (a) sandwich assays, in which unlabeled analytes are detected by a matched pair of antibodies specific for every protein target; and (b) label-based detection, in which the protein analyte is directly labeled by covalent attachment of tags, such as biotin or fluorophores (for example, Cy3 and Cy5), to enable detection after each targeted protein binds to the array. Sandwich assays can provide both high sensitivity and high specificity, and have been effectively demonstrated in the parallel measurements of low-abundance cytokines in culture supernatants and body fluids ***(3,10)***. Label-based

From: *Methods in Molecular Biology, Vol. 328: New and Emerging Proteomic Techniques*
Edited by: D. Nedelkov and R. W. Nelson © Humana Press Inc., Totowa, NJ

detection is an attractive complementary alternative to the sandwich assay, and its major advantage is ease in assay development, because only one antibody per target is required. Importantly, multicolor fluorescence detection is made possible when the targeted proteins are labeled. Because different samples may be labeled with different tags, a reference sample may be co-incubated with a test sample to provide internal normalization to account for concentration differences between spots. Although label-based detection is accurate and reproducible in the analysis of higher-abundance proteins, the detection sensitivity has not been sufficient to reliably detect lower-abundance proteins in biological samples using current methodology. Rolling-circle amplification (RCA) has been used for sensitivity enhancement in DNA quantitation ***(12)***, DNA mutation detection ***(13,14)***, and array-based sandwich immunoassays ***(3,15)***. RCA is well suited for multiplexed assays on solid surfaces, because the covalently attached amplified DNA cannot diffuse away. The isothermal process used in RCA takes place at moderate temperatures, and preserves the integrity of antibody–antigen complexes. RCA has been adapted to microarray-format immunoassays employing sandwich designs, as well as label-based detection. A schematic diagram illustrating alternative designs for antibody microarray assays utilizing RCA signal enhancement is shown in **Fig. 1**. In this chapter, we will focus on methods for the implementation of label-based assays, which are the most straightforward in implementation.

2. Materials

2.1. Microarray Preparation

1. Glass slides coated with a polyacrylamide hydrogel (hydrogel slides) (PerkinElmer Life Sciences, Boston, MA).
2. Antibodies, preferably of documented high specificity (*see* **Note 1**), are obtained from various commercial suppliers or from collaborators.
3. Aluminum foil tape #425–3 (R. S. Hughes Company, Sunnyvale, CA).

2.2. Protein and DNA Labeling

1. Amino-reactive fluorescent dyes, comprising *N*-hydroxysuccinimide (NHS) ester-linked Cy3 or Cy5 (Amersham Biosciences, Piscataway, NJ).
2. Amino-reactive haptens, NHS-digoxigenin, and NHS-biotin (Molecular Probes, Eugene, OR).
3. Chromatography spin columns with a molecular-weight cutoff of 6000 Da (Bio-spin P6) (Bio-Rad Laboratories, Hercules, CA).
4. Phosphoramidites coupled covalently to Cy3 or Cy5 dyes, for synthesis of labeled oligonucleotides (Glen Research, Sterling, VA).

2.3. RCA, Electrophoresis, and Microarray Washing Reagents

1. Phi 29 DNA polymerase (New England Biolabs, Beverly, MA).
2. Tango buffer (Fermentas, Hanover, NH).

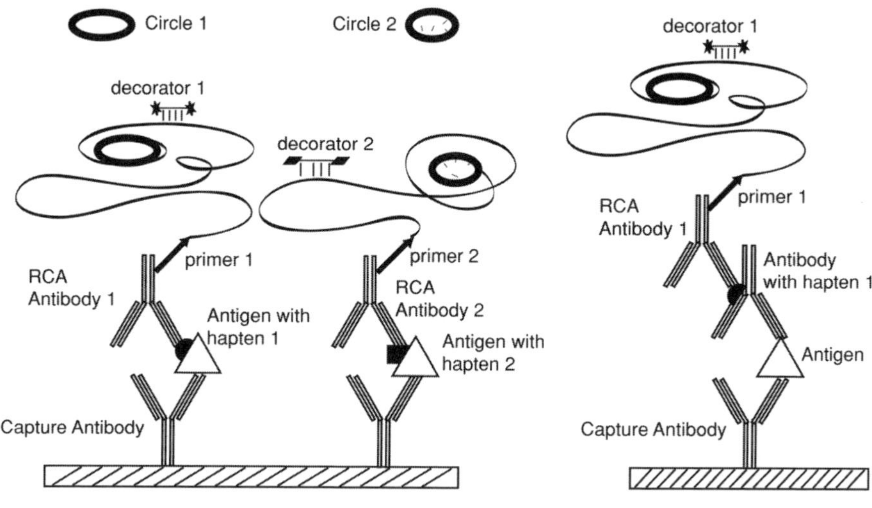

Fig. 1. Alternative designs for detection of antigens on the surface of antibody microarrays. In the label-based system employing haptenated antigens (diagram on the left) a capture antibody binds an antigen (labeled with hapten 1 or hapten 2). Preferred haptens are biotin and digoxygenin. The captured antigen is then detected by binding of an antibody-primer conjugate (rolling circle amplification [RCA] antibody 1 or 2), which generates amplified DNA after RCA, followed by fluorescent labeling of the amplified DNA with decorator 1 or decorator 2. A more elaborate assay system, based on a sandwich design, is depicted in the diagram on the right side. A capture antibody printed on a microarray binds an unlabeled antigen. The antigen is then bound by a haptenated second antibody (antibody with hapten 1) that recognizes a second epitope on the antigen. A third antibody, coupled to an oligonucleotide primer, binds to hapten 1 and functions as the signal generator. RCA is used to generate amplified DNA, followed by fluorescent labeling with decorator 1. Note that the sandwich system requires a matched pair of antibodies specific for every protein target, which insures extremely high specificity at the cost of added complexity. Because this system employs an unlabeled antigen, it is not straightforward to perform simultaneous two-color microarray experiments, and for this reason most investigators use a single-channel (single-color) detection system for sandwich assays. In this chapter, we describe methods for the label-based system based on haptenated antigens, which is simpler to implement and permits higher multiplexing.

3. TBE buffer: a 1X TBE solution contains 0.089 *M* Tris base, 0.089 *M* boric acid (pH 8.3), and 2 m*M* Na_2 ethylenediamine tetraacetic acid (EDTA).
4. PBST0.1 buffer: phosphate-buffered saline (PBS) with 0.1% Tween-20. Prepare by dissolving 8 g of NaCl, 0.2 g of KCl, 1.44 g of Na_2HPO_4, and 0.24 g of KH_2PO_4

in 800 mL distilled H_2O. Adjust pH to 7.4 with HCl, add 1 mL of Tween-20, and adjust volume to 1 L with additional distilled H_2O. Sterilize by autoclaving.

5. 2X sodium saline citrate (SSC) or 1X SSC. 1X SSCis 0.15 *M* NaCl, 0.015 *M* Na citrate.

2.4. DNA Primers for RCA, To Be Conjugated to Antibodies

1. The oligonucleotide p1, thiol-5′-CACAGCTGAGGATAGGACsAsT serves as the first RCA primer. The symbol "s" is used to indicate the presence of a phosphorothioate linkage in the last two residues proximal to the 3′-end of the oligonucleotide, which serve to protect from degradation by 3′-5′ exonuclease activity. The 5′-thiol group enables covalent coupling to antibodies. 5′-Thiol and phosphorothioate phosphoramidites for oligonucleotide synthesis are obtained from Glen Research, Sterling, VA.
2. The oligonucleotide p2, thiol-5′-TGTCTCAGTAGCTCGTCAsGsT serves as the second RCA primer. This primer (p2) was named primer 4.2 in publications from the Haab laboratory.

2.5. Preparation of DNA Circles

1. Circularization of oligonucleotide template for primer p1: The oligonucleotide c1, pGTCAGAACTCACCTGTTAGAAACTGTGAAGATCGCTTATTATGTCC-TATCCTCAGCTGTGTAACAACATGAAGATTGTAG contains a 5′-phosphate (the terminal-phosphate phosphoramidite is obtained from Glen Research) to enable circularization with DNA ligase. The oligonucleotide g1, GTGAGTTCT-GACCTACAATCTTCA-amino, which has a 3′-amino group to prevent priming by DNA polymerase (3′-amino phosphoramidite, Glen Research) serves as a guide for circularization. A 2 mL ligation reaction is prepared by mixing 200 µL of oligonucleotide c1 (12 µ*M*), 200 µL of oligonucleotide g1 (16 µ*M*) 200 µL of 10X ligation buffer, and 1348 µL of distilled water, followed by vortexing and heating to 65°C for 2 min. Then ligation is begun by adding 40 µL of bovine serum albumin (5 mg/mL stock) and 12.5 µL of T4 DNA ligase (400 U/mL), followed by incubation at 37°C for 2 h. DNA ligase is inactivated by incubation at 90°C for 2 min (*see* **Note 2**).
2. Circularization of oligonucleotide template for primer p2: The oligonucleotide c2, pCTGTGAGGTACTACCCTAATCGGACCTGTGAGGTACTACCCTAACT-TACTGACGAGCTACTGAGACATGTACAATCGGAC contains a 5′-phosphate (Glen Research) to enable circularization with DNA ligase. The oligonucleotide g2, AGTACCTCACAGGTCCGATTGTAC-amino, which has a 3′-amino group (Glen Research) serves as a guide for circularization. Circularization is performed as in **Subheading 2.5., step 1**.
3. Purification of circularized DNA. The reactions containing ligated circles are split into three microcentrifuce tubes (1.5-mL capacity) and deproteinized with an equal volume of a mixture of phenol:chlroroform (1:1) by vigorous shaking (15 s) in a vortex mixer. The aqueous phase (upper layer) is recovered after centrifugation for 30 s at 7000*g*. The aqueous phase is washed with an equal volume of chloroform,

centrifuged, and washed with chloroform a second time. The aqueous layer is now brought up to 0.3 *M* salt by addition of one-tenth volume of 3 *M* Na acetate (pH 5.2). DNA is precipitated with 2.5 volumes of 100% ethanol. After addition of ethanol, the tubes are vortexed vigorously and then placed at –20°C for at least 1 h (or left in the freezer overnight). DNA is recovered by centrifugation for 15 min at 12,000 RPM in a benchtop microcentrifuge. The pellet is washed with 200 μL of 80% ethanol and again recovered by centrifugation. Finally, the pellet containing circular DNA is dissolved in 85 μL of 10 m*M* Tris-HCl (pH 7.7), 0.1 m*M* EDTA. The three aliquots of circular DNA are pooled, yielding a total volume of 255 μL. Confirmation of circularization is by electrophoresis in denaturing polyacrylamide gels. A denaturing 10% polyacrylamide gel is prepared in Tris-borate-EDTA buffer containing 8 *M* urea. One microliter of circularized DNA is mixed with 7 μL of gel loading buffer (0.5X TBE, 10 *M* urea), and denatured for 2 min in a heat block set to 92°C. As a control, a mixture of all DNAs and reagents used for ligation, minus DNA ligase, is prepared and mixed with gel load buffer, to serve as an indicator of the electrophoretic mobility of noncircular DNA. The gel is run for 25 min at 24 volts/cm. The gel is stained for 35 min with SYBR Green II (Molecular Probes, Eugene, OR), which is used at a 1/10,000 dilution, as recommended by the manufacturer, and then photographed with UV illumination. Successful generation of circular DNA will result in the presence of a band of considerably lower mobility than the linear form of the 80-base c1 or c2 oligonucleotide. In general, the circular molecules visible in the gel are present as a strong 80-mer, a fairly strong 160-mer, and a weaker 240-mer. The dimeric and trimeric circular concatemers work just as well as the monomeric circle as templates for RCA.

2.6. Antibody(Ab)-Primer Covalent Adducts

1. Preparation of covalent adducts is based on conjugation of thiol-labeled DNA oligonucleotides (sequences specified under **Subheading 2.4.**) to reactive amino groups in antibodies, using a heterobifunctional cross-linker, essentially as described by Schweitzer et al. ***(15)***. Anti-biotin antibody is obtained from Zymed (mouse monoclonal 03–3700, Zymed Laboratories, San Francisco, CA) and covalently conjugated to a 20-base oligonucleotide primer (p1) containing a 5′-thiol group, as follows. Desalted Ab (41 nmoles) is treated with a 10-fold molar excess of sulfo-GMBS (Pierce Chemicals, Dallas, TX) under nitrogen in the dark for 30 min at 37°C, followed by 30 min at room temperature. Unreacted sulfo-GMBS is removed by chromatography over a PD-10 column (Bio-Rad Laboratories, Hercules, CA) equilibrated with sodium phosphate (pH 7.5)/150 m*M* NaCl. The Ab is then concentrated in a Centricon YM-30 (Millipore Corp., Billerica, MA) at 4°C. The number of maleimides per Ab is determined by utilizing Ellman's reagent (Pierce) to measure sulfhydryls following titration of β-mercaptoethanol by the activated Ab. An amount equal to 28.1 nmol of sulfo-GMBS-activated Ab and 142 nmol of 5′ thiol oligonucleotide is conjugated in a volume of 825 mL for 2 h at room temperature, followed by overnight at 4°C. Ab conjugated to oligonucleotide is purified by anion exchange chromatography on Q-Sepharose (Amersham Pharmacia) using a

salt gradient. Fractions containing conjugate are pooled and subjected to size exclusion chromatography on Superdex-200 (Amersham Biosciences-GE Healthcare, Piscataway, NJ) at 4°C to remove free oligonucleotide. A 10:1 ratio of oligonucleotide/Ab is typically used in conjugations, which results in a more reproducible production of conjugates, with three to five oligonucleotides per Ab molecule. Size exclusion chromatography is used to remove unconjugated oligonucleotide, ensuring a final preparation that is free of unconjugated Ab or DNA. This conjugation procedure yields a recovery of approx 50% of the starting Ab. Linear RCA reactions in solution ***(13)*** can be carried out to assess the ability of the Ab-DNA conjugate to serve as a primer. Standard RCA reactions contain 5 n*M* primer or Ab-DNA conjugate, 10 n*M* circular DNA template, and 0.4 U/μL of Phi29 DNA polymerase (New England Biolabs), 0.4 m*M* each dATP, dTTP, dGTP, and dCTP in 25 μL of Tango buffer (Fermentas). Additions are performed on ice and then shifted to 31°C. RCA products generated after 10, 20, and 30 min of incubation are assayed by electrophoresis for 2 h at 2.5 V/cm in a standard 0.7% agarose gel. The gel is stained for 35 min with Sybergreen II (Molecular Probes, Eugene, OR), which is used at a 1/10,000 dilution, as recommended by the manufacturer, and then photographed with UV illumination.
2. Anti-digoxygenin Ab covalently conjugated to a 20-base oligonucleotide primer (p2). Methods for generation of this adduct are identical to **Subheading 2.6. step 1**, except for the use of an anti-digoxygenin antibody (sheep anti-DIG IgG, Anawa Biomedical Services and Products, Switzerland) and a DNA oligonucleotide of different sequence.

2.7. Fluorescent "Decorator" Oligonucleotides

1. The oligonucleotide d1, cy3-ATGTCCTATCCTCAGCTGTG is used for fluorescent "decoration" of the amplified DNA generated by the RCA reaction on template c1.
2. The oligonucleotide d2, cy3-ACTGACGAGCTACTGAGACA is used for fluorescent "decoration" of the amplified DNA generated by the RCA reaction on template c2.

3. Methods

3.1. Printing of Antibody Microarrays

1. Antibody processing. The first step in producing antibody microarrays is the preparation of a set of antibodies. Antibodies work best in the microarray assay when they are highly purified. A high concentration of other proteins in the antibody solution usually results in a weakened or nonspecific signal, because binding sites on the microarray are occupied by the other proteins. Polyclonal antibodies collected from antisera should be antigen-affinity purified. Monoclonal antibodies that are provided in ascites fluid should be further purified using a kit such as the Bio-Rad Affi-gel Protein A MAPS kit. If bovine serum albumin (BSA) or gelatin has been added by the manufacturer, the IgG portion of the solution should be isolated

using Protein A purification. Glycerol is sometimes added to antibody solutions to prevent freezing at –20°C. Although glycerol will not interfere with the assay, the added viscosity may negatively affect the printing process. Glycerol concentrations above approx 20% should be avoided. To remove glycerol or to change the buffer of an antibody, we recommend the Bio-Rad Micro Bio-Spin P30 column (*see* **Note 3**). If the antibody is subsequently to be labeled, do not put the antibody in a Tris or amine-containing buffer, which will interfere with primary amine-based labeling reaction. The concentration of the antibody solutions should be adjusted to approx 500 µg/mL. Lower concentrations may also work well for some antibodies, and higher concentrations could yield higher signal intensities and may be desirable if consumption of antibody is not a concern.

2. Assembly of print plates. After the antibodies have been prepared at the proper purity and concentration, they are assembled into a "print plate"—a microtiter plate used in the robotic printing of the microarrays. Polypropylene microtiter plates are preferable to polystyrene because of lower protein adsorption. The plate should be rigid and precisely machined for optimal functioning with printing robots. The amount of antibody solution to load into each well of the print plate depends on the requirements of the printing robot—usually 10–15 µL is sufficient. If printing is sometimes inconsistent or variable between printing pins, it is desirable to fill multiple wells with the same antibody solution, so that different printing pins spot the same antibody and the data from replicate spots can be averaged. Store the 384-well print plates sealed in the refrigerator until ready to use. Aluminum foil tape provides a good seal (see Materials). Long-term evaporation-free storage is ensured by enclosing the covered plate in a sealed plastic bag (*see* **Note 4**). Prepare a spreadsheet containing the well identities for use in downstream data-processing applications.
3. Printing substrates. Various substrates for antibody microarrays have been demonstrated, such as poly-l-lysine-coated glass ***(1)***, aldehyde-coated glass ***(2)***, nitrocellulose ***(4,16)***, and a polyacrylamide-based hydrogel ***(17,18)***. Microscope slides with these various coatings can be purchased commercially. Because each application is unique in some aspects, the choice of which to use should be determined empirically by each user. We recommend simply preparing arrays on several different substrates and running the arrays in parallel, using the same sample on each array. Several criteria could be used to determine which surface type is best. The user should look at the signal-to-background ratio at each antibody spot, the reproducibility between replicate arrays, consistency in spot morphology, and the consistency in the background within each array.
4. Microarray printing hardware. Several varieties of commercial microarray printers are available, either contact printers or noncontact printers. Contact printers, commonly used for printing cDNA arrays, use metal pins to pick up antibody solutions from the print plate and deposit a small amount of the solutions (usually approx 1 nL) onto microarray substrates by contacting the surface. Noncontact printers deposit solutions without contacting the substrate, usually by ejection of a subnanoliter droplet from a quartz tip by a piezo-electric mechanism. Contact printers

can produce arrays faster than noncontact printers, especially if multiple printing pins are used simultaneously. The low production capacity of the noncontact printers can be a problem for some users, but the advantages of better spot reproducibility and morphology can be significant. Also, noncontact printers usually allow completely flexible array configurations, whereas the choices are usually more constrained using contact printers. The ability to print multiple replicate arrays on a single slide can be very useful, because the microarray substrates are used more efficiently and the throughput of the experiments can be higher. If multiple arrays per slide are printed, a method to separate the arrays is needed, so that different samples can be incubated on each array. A simple method is to draw a hydrophobic border around each array with a hydrophobic marker (*see* **Subheading 2.**). Another method for that purpose is the ProPlate (Grace Biolabs, Bend, OR), which can form up to 16 separate "wells" around 16 different microarrays on each slide.

5. Microarray printing. The details of the printing process will depend on the type of printing robot used, but we give some general notes here. Minimize the time that the print plates are unsealed and exposed in order to keep evaporation of the antibody solutions low. Evaporation may be minimized by cooling the print plate and maintaining an appropriate humidity in the printing environment. In a room-temperature environment, cooling the print plate to 5°C at a humidity of approx 43% will minimize both evaporation and condensation. The user should confirm that the print tips are sufficiently washed between loads, so that no measurable contamination between antibodies occurs. This test can be performed by printing a dye-labeled protein solution and a buffer solution in successive spots. If fluorescence is seen in the buffer spots, the pins need to be washed more stringently. Most microarrayers will allow the printing of replicate spots on each array, which are useful to obtain more precise data through averaging and to ensure the acquisition of data if a portion of the array is unusable. Three to six spots per array per antibody are usually sufficient. The handling of the arrays after printing depends on the surface used—see the manufacturer recommendations. Usually it will be advised to let the arrays incubate for some time to allow the antibodies to fully bind to the substrate. We recommend vacuum sealing and refrigerating the arrays for storage to minimize loss of antibody activity (*see* **Note 5**).

3.2. Sample Labeling With Haptens

1. Hapten-labeled protein samples are prepared by covalent labeling of an aliquot of sample with NHS-digoxigenin, and another aliquot with NHS-biotin. Each protein sample aliquot is diluted 1:15 with 50 m*M* carbonate buffer at pH 8.3, followed by addition of 1/20 volume of 6.7 m*M* NHS ester linked to digoxigenin or biotin, dissolved in dimethylsulfoxide (DMSO). After the reactions proceed for 1 h on ice, one-tenth volume of 1 *M* Tris-HCl (pH 8.0) is added to each tube to quench the reactions, and the solutions are allowed to sit for another 20 min. The unreacted dye is removed by passing each solution through a size-exclusion chromatography spin column (Bio-Spin P6). The digoxigenin-labeled samples are pooled, and equal

amounts of the pool are transferred to each of the biotin-labeled samples. Each hapten-labeled protein solution is supplemented with nonfat milk to a final concentration of 3%, Tween-20 to a final concentration of 0.1%, and 1X PBS to the final desired dilution. The optimal dilution of the samples may vary depending on the surface type and antibodies used, and should be determined empirically.

3.3. Microarray Immunoassay and RCA Signal Generation

The following reagents, described under **Subheading 2.**, are used: Anti-biotin antibody covalently conjugated to a 20-base oligonucleotide p1; anti-digoxigenin antibody covalently conjugated to a different 20-base oligonucleotide p2; 80-base circular DNA c1 with a portion complementary to primer 1; 80-base circular DNA c2 with a portion complementary to primer p2; Phi29 DNA polymerase; and decorator oligonucleotides d1 and d2.

1. Contacting of labeled protein sample with antibody microarrays: A 100-μL sample of each hapten-labeled serum sample mix is incubated on a microarray with gentle rocking at room temperature for 1 h. The microarrays are rinsed briefly in PBST0.1 to remove the sample, washed three times for 3 min each in PBST0.1, and dried by centrifugation. (*Optional:* for indirect detection *without* RCA, microarrays are incubated under gentle rocking for 1 h at room temperature with primer 1-labeled anti-biotin and primer 2-labeled anti-digoxigenin antibodies, each prepared at 0.5 μg/mL in 1X PBST0.1 with 1% BSA. A Cy3-labeled 20-bp oligonucleotide (d1) complementary to primer 1 and a Cy5-labeled 20-bp oligonucleotide (d2) complementary to primer 2 are prepared at 0.2 m*M* each in 2X SSC with 0.1% Tween-20 and 0.5 mg/mL herring sperm DNA. This solution is incubated on the microarrays for 1 h at 37°C. By using these labeled "decorator" oligos, d1 and d2, the antibodies are labeled by hybridization to the covalently bound primers.)
2. For RCA-enhanced detection, the microarrays are incubated for 1 h at room temperature with a solution containing 75 n*M* oligonucleotide c1, 75 n*M* oligonucleotide c2, 1.0 μg/mL primer p1-conjugated antibiotin, and 1.0 μg/mL primer p2-conjugated anti-digoxigenin in PBST0.1 with 1 m*M* EDTA and 5 mg/mL BSA. The microarrays are rinsed briefly in PBST0.1 and washed at room temperature with gentle rocking three times for 3 min each in PBST0.1. Phi29 DNA polymerase (Biolabs) in 100 μL of 1X Tango buffer (Fermentas) solution with 0.1% Tween-20 and 1 m*M* dNTPs is incubated on the microarrays at 37°C for 30 min. The final concentration of Phi29 DNA polymerase should be 0.4 U/μL. After RCA, the microarrays are rinsed briefly in 2X SSC/0.1% Tween-20, washed three times for 3 min each at room temperature with gentle rocking in 2X SSC/0.1% Tween-20, and dried by centrifugation. A Cy3-labeled 20-bp decorator oligonucleotide (d1) complementary to the repeating DNA strand generated from primer p1 and a Cy5-labeled 20-bp decorator oligonucleotide (d2) complementary to the repeating DNA strand from primer p2 are prepared at 0.2 m*M* each in 2X SSC with 0.1% Tween-20 and 0.5 mg/mL herring sperm DNA. This solution is incubated on the microarrays for 1 h at 37°C, followed by washing for 2 min in 1X SSC.

3.4. Scanning of Microarrays

Several commercial microarray scanners are available, and each will have slightly different capabilities and functionality. Features such as adjustable resolutions, adjustable laser and photomultiplier settings, and autofocus abilities may vary between scanners. Some have the option to scan more than two channels of fluorescence, or are equipped with autoloaders for the convenient scanning of many slides (*see* **Note 6**). That capability is especially valuable for high-resolution scans of multiple slides.

1. Scanner settings. Most currently available scanners allow the user to adjust the laser and photomultiplier tube (PMT) settings, which is necessary because signal strengths can vary significantly between experiments, depending on the surface and sample types and other factors. The following guidelines should be adhered to. A significant saturation of signal at any of the spots should be avoided, because the values for those spots cannot be known if the signal is saturated (*see* **Note 7**). Therefore, the settings should be as high as possible without saturating any spots; usually, the average spots reach about 50% of the detector capacity. Most scanners offer a "quick scan" option at low resolution, to allow the user to rapidly determine the signal levels on the array and to set the optimal scanner settings. The laser power should be preferentially increased over the PMT gain. Higher laser power can improve signal-to-noise ratios ***(19)***, and laser outputs are more stable (less variable) at higher powers. The laser should be set as high as possible—up to approx 90–95% of maximum power. If the laser power is maximized and more signal is desired, increase the PMT setting. When scanning a group of slides that, together, compose an experiment set, it is usually desirable to use the same scanner settings for all slides. The user may then perform a quick scan on each slide in the set to verify that all the slides may be scanned at the chosen setting. The prescans should be limited to minimize photobleaching of the fluorophores on the arrays.
2. After the laser and PMT settings have been chosen, scan the slides at a resolution appropriate to the spot sizes. Ten to twenty pixels across the diameter of each spot are optimal. So if the spots have a diameter of 200 μ*M*, use a resolution of no more that 20 μ*M*. Fewer than 10 pixels across a spot can lead to imprecise averaging, and more than 20 pixels across does not further improve data quality and results in unnecessarily large image file sizes. Most scanners save the images in the TIFF format, with a separate image for each color channel that was scanned. An image of a scanned sub-array is shown in **Fig. 2**.

3.5. Data Analysis

Data analysis involves quantification of fluorescence intensity, spot flagging, averaging of replicate spots, normalization, and statistical analysis. The logical flow of these steps is outlined as follows, but a full treatment of the complex subject of microarray data analysis is beyond the scope of this chapter. The reader is referred to recent reviews ***(21–23)*** of the extensive literature on the subject.

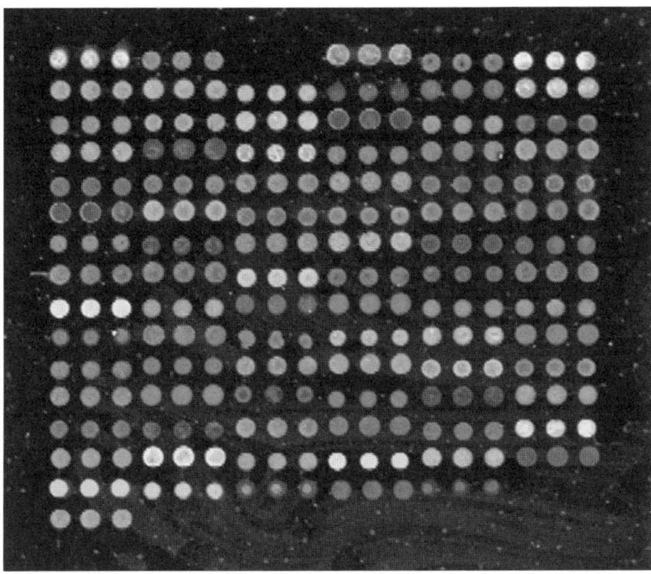

Fig. 2. Scanned image of an antibody sub-array. Antibody replication is in triplicate.

1. Spot localization and quantification of fluorescence intensity. The data analysis begins with quantification of the scanned images using a microarray gridding program, such as GenePix Pro from Axon Laboratories. This program will assign user-supplied identifications to each spot, locate the boundaries of each spot, and calculate various statistics for each spot, such as the average signal intensity of the pixels within a spot and outside of a spot, or the standard deviation between the pixels. The use of these numbers then depends on the type and purpose of the experiments. Two-color, label-based assays have significantly different primary data analysis methods ***(7,16)*** as compared with single-color sandwich assays ***(20)***.
2. Low-level data processing. The primary processing of the data involves spot flagging, averaging of replicate spots, and normalization. Spot flagging is the acceptance or rejection of spots based on signal intensities or other parameters. All arrays should be visually inspected, and spots with obvious defects should be flagged manually. A useful threshold for flagging spots by intensity is calculated by the formula $3*B*CV_b$, where B is each spot's median local background, and CV_b is the average coefficient of variation (standard deviation divided by the average) of all the local backgrounds on the array. This threshold is similar to the standard deviation of the local background but minimizes the effects from spikes in the local backgrounds. For two-color data, the relevant parameter for subsequent analysis is the ratio of the fluorescence from the two color channels. For each spot, use the formula (F1–B1)/(F2–B2), where F1 and F2 are the median intensities of the pixels inside the spot for color channels 1 and 2, respectively, and B1 and B2 are the

median intensities of the pixels locally outside the spot for the two color channels. GenePix Pro calculates this parameter automatically, and the user can simply extract the values of the nonflagged spots. The averaging of replicate spots helps to reduce variability caused by spot-to-spot differences. The "geometric mean" of the ratios should be calculated. That is, the logarithm of the ratios should be calculated (base 2 is useful) prior to averaging. The averages can then be anti-logged or left as logs, depending on the next analysis.

3. Normalization is typically applied to microarray data to account for possible systematic experimental variation in factors such as sample labeling efficiency, scanner readout efficiency, and microarray quality ***(24,25)***. Methods developed for two-color cDNA array data may be useful here, but the differences in antibody microarray experiments, such as a smaller and more selected set of targets and a different labeling method, may mean that the optimal normalization methods may be different. A simple "global normalization" method, which sets the median or mean ratio on each array to a fixed value, usually works quite well, but that method relies on the assumption that the average abundance of the measured proteins does not change between samples. A better method might be to select a subset of the spots within each array—spots that are expected not to change between samples—and to set the mean of those spots equal between arrays. If that method is used, the arrays should be designed to include many common, housekeeping, or structural proteins for use in the normalization.
4. Once the data are normalized, data from an experiment set can be brought together for statistical analyses. A variety of well-documented microarray analysis packages are available (microarrayworld.com/SoftwarePage.html), and the most appropriate tools will depend on the goals of the experiment. A common goal is to identify differences between samples, or to test the use of groups measurements for the classification of sample types. Often it is useful simply to visualize the expression profiles in a data set. Clustering is a useful tool for that, and a widely used program is freely available (rana.lbl.gov). Clustering allows the convenient visualization of patterns from many proteins over many samples and also the assessment of the degree of similarity between samples or proteins.

4. Notes

1. Antibodies should be of very high specificity, as determined by enzyme-linked immunosorbent assay (ELISA) or other methods. Specificity can be tested on the microarray format by performing competition assays, provided that the intended antigen is readily available. Serum is spiked with antigen, and labeling is performed as indicated. Increasing amounts of unlabeled antigen are added in a series of microarray experiments, to measure the reduction of the fluorescent signal as a function of the concentration of competing antigen.
2. Exonuclease digestion after generation of circular DNA was used in previously published protocols to destroy the remaining guide oligonucleotides after ligation. However, we have eliminated this step by using guide oligonucleotides with blocked 3′ termini, which have no priming activity. An amino group located at the

3′ end of the guide oligonucleotide can serve this purpose. The guide oligonucleotides remain hybridized to the circular oligos during purification, but are readily displaced by phi29 DNA polymerase during the RCA reaction.

3. The Biospin columns come prepacked with two types of buffers: SSC and Tris buffer. The packing buffer comes out of the column with the sample that was applied to it. That is, after a sample is run through the column, it will be in the buffer with which the column was packed. The packing buffer can be changed by running a different buffer through the column three times. The P30 column removes solution components smaller than 30 kD, and the P6 column removes components smaller than 6KD. Thus the P30 column is better for purification of antibodies, and the P6 column is better for the purification of complex mixtures in which low-molecular-weight species should be preserved.
4. Food storage sealers work well for this purpose. Some models have the option of applying vacuum, or simply sealing without vacuum. To completely prevent evaporation of fluid from the print plate, insert a moist piece of paper towel into the bag with the print plate before sealing; this will keep the humidity in the bag high.
5. We do not know of systematic studies on the shelf life of antibody microarrays. Our own experience has show that arrays stored up to 2 mo under vacuum in a refrigerator or cold-room show undiminished signal and performance properties.
6. A cautionary note when using autoloaders is that high levels of atmospheric ozone can be destructive to certain dyes, such as Cy5 ***(26)***. High levels of ozone can build up in closed compartments with electrical motors inside.
7. Saturation occurs when the upper limit of quantification of a signal has been exceeded. Most scanners will have some indication of pixels that have been saturated, such as white-colored pixels.

Acknowledgments

The authors thank Patrick O. Brown and David C. Ward for their enthusiastic support during the early stages of development of antibody microarrays and RCA technology. This work was supported by the Early Detection Research Network of the National Cancer Institute (grant U01 CA084986 to B.B.H.), the Michigan Proteome Consortium and the Michigan Economic Development Corporation (to B.B.H.), the National Cancer Institute (grant CA81671 to P.L.), a pilot project grant from the Yale Liver Center (grant P30 DK 34989 to P.L.), and the Van Andel Research Institute (to B.B.H.).

References

1. Haab, B. B., Dunham, M. J., and Brown, P. O. (2001) Protein microarrays for highly parallel detection and quantitation of specific proteins and antibodies in complex solutions. *Genome Biol.* **2,** 1–13.
2. MacBeath, G. and Schreiber, S. L. (2000) Printing proteins as microarrays for high-throughput function determination. *Science* **289,** 1760–1763.

3. Schweitzer, B., Roberts, S., Grimwade, B., et al. (2002) Multiplexed protein profiling on microarrays by rolling-circle amplification. *Nat. Biotechnol.* **20,** 359–365.
4. Knezevic, V., Leethanakul, C., Bichsel, V. E., et al. (2001) Proteomic profiling of the cancer microenvironment by antibody arrays. *Proteomics* **1,** 1271–1278.
5. Sreekumar, A., Nyati, M. K., Varambally, S., et al. (2001) Profiling of cancer cells using protein microarrays: discovery of novel radiation-regulated proteins. *Cancer Res.* **61,** 7585–7593.
6. Huang, R-P., Huang, R., Fan, Y., and Lin, Y. (2001) Simultaneous detection of multiple cytokines from conditioned media and patient's sera by an antibody-based protein array system. *Anal. Biochem.* **294,** 55–62.
7. Miller, J. C., Zhou, H., Kwekel, J., et al. (2003) Antibody microarray profiling of human prostate cancer sera: antibody screening and identification of potential biomarkers. *Proteomics* **3,** 56–63.
8. Nielsen, U. B., Cardone, M. H., Sinskey, A. J., MacBeath, G., and Sorger, P. K. (2003) Profiling receptor tyrosine kinase activation by using Ab microarrays. *Proc. Natl. Acad. Sci. USA* **100,** 9330–9335.
9. Huang, R., Lin, Y., Wang, C. C., et al. (2002) Connexin 43 suppresses human glioblastoma cell growth by down-regulation of monocyte chemotactic protein 1, as discovered using protein array technology. *Cancer Res.* **62,** 2806–2812.
10. Lin, Y., Huang, R., Santanam, N., Liu, Y. G., Parthasarathy, S., and Huang, R. P. (2002) Profiling of human cytokines in healthy individuals with vitamin E supplementation by antibody array. *Cancer Lett.* **187,** 17–24.
11. Tannapfel, A., Anhalt, K., Hausermann, P., et al. (2003) Identification of novel proteins associated with hepatocellular carcinomas using protein microarrays. *J. Pathol.* **201,** 238–249.
12. Nallur, G., Luo, C., Fang, L., et al. (2001) Signal amplification by rolling circle amplification on DNA microarrays. *Nucleic Acids Res.* **29,** E118.
13. Lizardi, P. M., Huang, X., Zhu, Z., Bray-Ward, P., Thomas, D. C., and Ward, D. C. (1998) Mutation detection and single-molecule counting using isothermal rolling-circle amplification. *Nat. Genet.* **19,** 225–232.
14. Ladner, D. P., Leamon, J. H., Hamann, S., et al. (2001) Multiplex detection of hotspot mutations by rolling circle enabled universal microarrays. *Lab. Invest.* **81,** 1079–1086.
15. Schweitzer, B., Wiltshire, S., Lambert, J., et al. (2000) Imunoassays with rolling circle DNA amplification: a versatile platform for ultrasensitive antigen detection. *Proc. Natl. Acad. Sci. USA* **97,** 10,113–10,119.
16. Zhou, H., Bouwman, K., Schotanus, M., et al. (2004) Two-color, rolling-circle amplification on antibody microarrays for sensitive, multiplexed serum-protein measurements. *Genome Biol.* **5**, R28.
17. Guschin, D., Yershov, G., Zaslavsky, A., et al. (1997) Manual manufacturing of oligonucleotide, DNA, and protein microchips. *Anal. Biochem.* **250**, 203–211.
18. Arenkov, P., Kukhtin, A., Gemmell, A., Voloshchuk, S., Chupeeva, V., and Mirzabekov, A. (2000) Protein microchips: Use for immunoassay and enzymatic reactions. *Anal. Biochem.* **278**, 123–131.

19. Mathies, R. A., Peck, K., and Stryer, L. (1990) Optimization of high-sensitivity fluorescence detection. *Anal. Chem.* **62**, 1786–91.
20. Nielsen, U. B. and Geierstanger, B. H. (2004) Multiplexed sandwich assays in microarray format. *J. Immunol. Methods* **290**, 107–120.
21. Leung, Y. F. and Cavalieri, D. (2003) Fundamentals of cDNA microarray data analysis. *Trends Genet.* **19(11),** 649–659.
22. Bolstad, B. M., Collin, F., Simpson, K. M., Irizarry, R. A., and Speed, T. P. (2004) Experimental design and low-level analysis of microarray data. *Int. Rev. Neurobiol.* **60,** 25–58.
23. Ochs, M. F., Moloshok, T. D., Bidaut, G., and Toby, G. (2004) Bayesian decomposition: analyzing microarray data within a biological context. *Ann. NY Acad. Sci.* **1020,** 212–226.
24. Park, T., Yi, S. G., Kang, S. H., Lee, S., Lee, Y. S., and Simon, R. (2003) Evaluation of normalization methods for microarray data. *BMC Bioinformatics* **4,** 33.
25. Yang, Y. H., Dudoit, S., Luu, P., Lin, D. M., Peng, V., Ngai, J., and Speed, T. P. (2002) Normalization for cDNA microarray data: a robust composite method addressing single and multiple slide systematic variation. *Nucleic Acids Res.* **30,** e15.
26. Fare, T. L., Coffey, E. M., Dai, H., et al. (2003) Effects of atmospheric ozone on microarray data quality. *Anal. Chem.* **75,** 4672–4675.

3

Antibody Microarrays Using Resonance Light-Scattering Particles for Detection

Bernhard H. Geierstanger, Petri Saviranta, and Achim Brinker

Summary

Antibody microarray measurements show great potential for the simultaneous quantification of many proteins in small amounts of body fluids and extracts. Over the last few years, a microarray platform centered around the concept of microarrays in microtiter wells was developed, and for the best assays we have achieved lower limits of detection in the femtomolar range using resonance light-scattering particles for the staining of biotinylated detection antibodies. Although conceptually simple, these multiplexed sandwich assays are technically challenging. Here we describe in detail our protocols and procedures for the manufacturing of antibody microarrays with up to 48 different antibodies and for performing plasma measurements.

Key Words: Antibody microarray; multiplexed sandwich assays; plasma proteins; resonance light-scattering (RLS) particles.

1. Introduction

Protein and antibody microarray measurements can be performed in a number of different ways and formats *(1–4)*. Reverse arrays wherein many samples, typically cell lysates *(5,6)* or tissue sections *(7)*, are probed simultaneously by a single antibody, are a powerful approach to characterize patient-derived, primary tumor samples, especially when employed in conjunction with laser capture microdissection. Reverse arrays have also been developed to study signal transduction pathways *(8)*. Arrays of antigens for the detection of auto-antibodies *(9–11)* have had the biggest impact so far in clinical environments. Two-color approaches that simultaneously compare the responses of large antibody arrays to chemically labeled reference and diseased samples continue to be powerful approaches to disease marker discovery *(3,12)*. Closest to the enzyme-linked immunosorbent assay (ELISA) tests widely used in research and clinical

From: *Methods in Molecular Biology, Vol. 328: New and Emerging Proteomic Techniques*
Edited by: D. Nedelkov and R. W. Nelson © Humana Press Inc., Totowa, NJ

laboratories are antibody microarrays using multiplexed sandwich assays. In this approach, different capture antibodies are spatially arrayed on a surface and a variety of detection schemes can be employed, as has been reviewed recently (*see* **refs. *13*** and ***14***). All achieve multiplexing capability by applying mixtures of detection antibodies to an array of immobilized capture antibodies ***(15–23)***. In this sense, antibody microarrays are simply multiplexed sandwich assays. Compared with a single assay in a microtiter plate, the concentrations and amounts of reagents involved are, however, quite different ***(24,25)***, and it is believed that for this reason, microarray measurements generally fall into the ambient analyte regime ***(26)***, meaning that the measurement does not deplete the analyte in the assay volume.

Because quantitative assays require calibration curves, we have developed a slide holder that divides a 1-inch × 3-inch slide into 48 separate wells (4 rows by 12 columns). The well-to-well spacing, 4.5 mm, is the same as is used in the standard 384-well microtiter plates. Each well contains the same 12 × 12 antibody microarray with 144 spots (triplicate spots for 48 antibodies; *see* **Note 1**). We use eight wells per slide to simultaneously generate eight-point calibration curves for all analytes, using mixtures of purified antigens. A single slide holder contains four slides with 48 identical antibody microarrays, and on each slide four dilutions of ten samples are measured against an eight-point calibration curve ***(25)***. Unique to our platform is the use of resonance light-scattering (RLS) particles, in essence 78-nm diameter colloidal gold particles coated with an anti-biotin antibody ***(27–29)***, for the detection of the biotinylated primary antibodies by signal generation through light scattering.

As we have shown recently, RLS can detect as few as 400 binding events on the surface of a 24-plex mouse antibody array ***(25)***. The light-scattering intensity of these colloidal gold particles is such that single particles can be seen under a light microscope ***(30)***. RLS staining has also been shown to dramatically increase the number of detectable gene messages in microarray gene expression profiling experiments ***(30)***. The high sensitivity of the stain is based on the light-scattering properties of colloidal gold particles ***(27–29)***, and others have used colloidal gold for the ultrasensitive, almost single-molecule detection of antigens after PCR amplification of DNA barcodes on nanoparticles ***(31)***. As will be illustrated below, RLS staining of biotinylated reagents is a fast, simple, and robust procedure that yields highly sensitive multiplexed sandwich assays. We routinely use RLS particles with two antibody arrays for human plasma proteins with 43 and 45 analytes respectively, as well as a 24-analyte antibody array for mouse plasma proteins. In addition, we are currently evaluating RLS staining procedures for microarray assays that study protein–DNA interactions as well as profiling of proteases using PNA-encoded probes ***(32)***.

The technical challenges of array measurements are numerous. With the RLS particles, we have found a very sensitive and robust staining method. Manufacturing of antibody arrays is a process that needs a strong commitment to details and tight process control, but good results can be obtained. Assay optimization, and availability and reliability of the reagents, will continue to be a challenge. In our experience, a maximum of 50 analytes can be multiplexed before cross-reactivity issues and antigen-independent antibody cross-talk will severely limit and affect the dynamic range and sensitivity of many sandwich assays on the array. There seems to be an intrinsic barrier for the degree of multiplexing. Needless to say, keeping a collection of 50 capture antibodies, 50 antigens, and 50 detection antibodies fully functional on a day-to-day basis is a formidable task. Smaller sets of reagents can be used over extended periods with good results, but clearly better antibody reagents are urgently needed. Because of the multiplexed nature of these measurements, the diluent and sample preparation methods cannot be optimized for best performance of all analytes. As has recently been shown ***(33)***, the choice of anticoagulant seems to have an effect on the results for specific analytes, and therefore compromises have to be made with respect to sample preparation and assay optimization.

2. Materials

2.1. Antibody Microarray Printing and Processing

1. GAPS II or UltraGAPS amino-silane coated glass microarray substrates are purchased from Corning, Inc. (*see* **Note 2**).
2. A Genemachine Accent printer from Genomic Solutions (Ann Arbor, MI) is equipped with a humidity control unit and an in-house built cooling unit for the print plate. A larger blotting pad is also installed (*see* **Note 3**).
3. Four SMP3 steel quill pins (Telechem, Sunnyvale, CA) are typically used in parallel.
4. Monoclonal and affinity purified polyclonal antibodies to be immobilized on the microarrays are purchased from a variety of vendors and used without further purification (*see* **Note 4**). Antibodies are stored according to the suppliers' recommendations, typically in liquid form at 4°C.
5. Capture antibodies are diluted to a final concentration of 0.25 mg/mL into a printing buffer that is optimized to the hydrophobicity of the surface as well as the printing conditions. The 5X printing buffer is available from Invitrogen (Carlsbad, CA).
6. Casein blocking solution (1%) for blocking of antibody arrays in preparation for storage is purchased from Pierce, and filtered through a 0.22-µm filter at room temperature. The filtrate is stored at 4°C.

2.2. Preparation and Storage of Samples

Complete, ethylenediamine tetraacetic acid (EDTA)-free protease inhibitor cocktail tablets are purchased from Roche, Inc.

2.3. Sandwich Microarray Assays

1. Biotinylated detection antibodies are purchased from a variety of vendors and used without further purification at concentrations between 0.013 and 0.5 μg/mL, as determined during assay optimization. Detection antibodies are stored according to the suppliers' recommendations, typically in liquid form at 4°C.
2. Other detection antibodies are biotinylated using NHS-PEO$_4$-biotin (Pierce) according to protocols provided by the manufacturer (*see* **Note 5**).
3. For the standard curves, antigens are purchased from a variety of vendors and used without further purification. Aliquots of the antigens are flash frozen in liquid nitrogen and stored at –80°C.
4. Antigens are diluted into casein blocking solution. Dilutions for calibration curves are freshly prepared for each assay from flash-frozen stock solutions (*see* **Note 6**). Casein blocking solution (1%) is purchased from Pierce.
5. Bovine serum albumin, sodium chloride, and Tween-20 are purchased from Sigma.
6. Goat IgG is purchased from Jackson ImmunoResearch Laboratories, Inc.
7. Phosphate-buffered saline (PBS) stock solution (10X PBS) is purchased from Invitrogen.
8. Complete, EDTA-free protease inhibitor cocktail tablets are purchased from Roche, Inc.
9. RLS particles and archiving solution are purchased from Invitrogen (Carlsbad, CA).
10. A slide holder to hold four microarray substrates was developed in house. The slide holder has 192 wells with 4.5-mm spacing separating each of the four microarray substrates into 48 physically separated 3 mm × 3 mm areas, similar to the footprint of a standard 384-well microtiter plate. The slide holder is commercially available through GNF's Department of Engineering.
11. A 384-plate washer (Tecan, Research Triangle Park, NC) was modified to allow for automated and simultaneous washing of all 192 wells in the slide holder. Generally, three cycles of 400 μL wash volume per well in the overflow wash mode are used (*see* **Note 7**). The wash solution consists of 0.005% Tween-20 in 1X PBS.
12. During all incubation steps the slide holder is shaken on a Tecan orbital shaker at 1500 rpm, changing directions (clockwise to counter-clockwise) every 5 min (*see* **Note 8**).

2.4. Microarray Imaging, and Image and Data Analysis

1. A 16-bit charge-coupled device (CCD) camera-based scanner optimized for the detection of resonance light scattering is available from Invitrogen (Carlsbad, CA).
2. Images are processed with ArrayVision, Version 8.0 (Imaging Research, St. Catharines, Canada). The software is provided with the scanner.
3. For quantification of the images, an Excel macro was developed. Details are described under **Subheading 3.4.** Sample files will be made available upon request.

2.5. Assay Development and Optimization

Reagents and equipment for the development of new assays were obtained as described previously.

3. Methods

Antibody microarray measurements as described here are conceptually simply multiplexed sandwich assays in microarray format. Rather than obtaining a concentration readout for a single protein of interest, tens of proteins are measured simultaneously with the same small amount of sample. In our format, 15 µL of sample are sufficient for the quantification of up to 45 different analytes (*see* **Note 1**). However, in practice, microarray measurements are quite complex. This starts with the manufacturing of microarrays that requires tight environmental and process control, and great care must be taken in optimizing printing conditions to the surface properties of the microarray substrate before acceptable results can be obtained (*see* **Subheading 3.1.**). Because of the multiplexed nature of the assay, compromises must be made in how samples can be handled (*see* **Subheading 3.2.**). The sandwich microarray assay (*see* **Subheading 3.3.**) itself roughly follows the scheme of an ELISA plate assay, with a sample incubation period, a detection antibody incubation period, and a staining step, with additional blocking and washing steps in between. Staining with RLS particles follows a simple procedure that results in a highly sensitive and generic signal generation method for biotinylated reagents frequently used in microarray applications, such as protein, antibody, oligonucleotide, and cDNA microarrays ***(30)***. In contrast with plate assays, microarray measurements require an image to be taken and analyzed (*see* **Subheading 3.4.**). We will also describe how we calibrate the assays and analyze the actual protein concentrations. Last but not least, the process of developing and optimizing a panel of assays into an antibody microarray is discussed under **Subheading 3.5.**, and likely constitutes the most challenging aspect of antibody microarray measurements. Processing of 16 microarray slides in four slide holders (with a total of 768 microarrays), from mounting the slides, performing the assay, to archiving all slides, requires approx 5 h.

3.1. Antibody Microarray Printing and Processing

3.1.1. Preparing Print Plate

1. To avoid transferring particulate matter to the print plate, capture antibody stock solutions are spun for 30 min at 30,000*g* (at 4°C).
2. Capture antibody stock solutions are diluted with 1X PBS and 5X printing buffer to a final antibody concentration of 0.25 mg/mL (*see* **Note 9**).

3. 8 μL of each antibody printing solution is transferred to a 384-Well Round Bottom Low Volume plate (Corning, part number 3676).
4. After all samples have been added, the print plate is sealed with an aluminum adhesive lid (Beckman) and spun at 500*g* for 30 s to remove air bubbles and to settle the samples to the bottom of the wells.
5. The print plate is stored in the cold-room for at least 24 h before the printing operation begins (*see* **Note 10**).

3.1.2. Preparing Microarrayer and Printing of Microarrays

1. The microarrayer is situated in a dedicated humidity- and temperature-controlled room that is supplied with high-efficiency particulate arresting (HEPA)-filtered air generating a near clean-room environment. The room is set to 70% relative humidity and 20°C. The humidity control on the microarrayer is set to 75% relative humidity (*see* **Note 11**).
2. The array printer is equipped with a temperature-controlled stage used to keep the print plate at a constant temperature of approx 15.5°C throughout the printing process. The temperature for the sample plates is set just slightly above the dew point to prevent evaporation and concentration of the samples to be printed as well as to prevent condensation (*see* **Note 12**).
3. SMP3 pins deposit approx 0.6 nL of solution in a single spot. 50 spots are preprinted onto a blotting pad to remove excess material from the outside of the pin. Antibodies are printed in triplicates with a spot-to-spot distance of 200 μm. 48 identical subarrays with 4.5-mm array-to-array spacing are printed with four pins in parallel. Each pin is refilled after 72 spots have been printed onto two microarray slides (two rows of 12 subarrays with triplicate spots per subarray) (*see* **Note 13**). The print plate is allowed to equilibrate to the stage temperature while basic printer maintenance, such as removing dust, changing rinse and cleaning solutions, and cleaning blotting pad, is performed.
4. Print pins are cleaned and inspected under microscope (*see* **Note 14**).
5. The sonication station is filled with fresh, deionized water (Millipore).
6. The print pins are inserted into the proper locations in the print head.
7. In the final step, the array substrates are mounted on the print deck, the cover is removed from the print plate, and the printing program is started.

3.1.3. Postprinting Processing and Storage

1. Microarrays are kept on printer for at least 1 h after printer has finished spotting the arrays.
2. To block printed slides with casein solution as preparation for storage, the antibody microarray slides are removed one by one from the printing deck, and gently and slowly lowered with a pair of forceps into a slide-mailer filled to three-quarters with blocking solution (*see* **Note 15**).
3. After 1 h room temperature incubation, remove slides from the slide-mailer one by one and place each slide in its own 50-mL centrifuge tube (Falcon tube).

4. To remove the excess blocking reagent from the slides, centrifuge the slides for 1 min at 285*g*.
5. Remove slides from the 50-mL centrifuge tubes and place them into a clean slide-mailer. Store slides at room temperature in a desiccator until they are needed (*see* **Note 16**).

3.2. Preparation and Storage of Samples

1. Our collaborators collected up to 150 μL of blood from mice via retro-orbital eye bleeds into Sarstedt Microvette 200 K-EDTA RB blood tubes (Fisher Scientific) or into heparin-coated collection tubes. Blood cells were separated by centrifugation, and plasma samples were flash frozen in liquid nitrogen. Similarly prepared human samples were provided to us by a number of collaborators in frozen form.
2. Samples are stored as frozen aliquots at –80°C.
3. After thawing, the samples are spun for 30 min at 30,000*g* at 4°C to remove precipitates and aggregates.
4. Complete, EDTA-free protease inhibitor cocktail is applied at a concentration of one tablet per 50 mL of diluent used throughout the assays.
5. For samples that are obtained in liquid form, protease inhibitor cocktail is added as soon as possible before flash freezing aliquots of samples.

3.3. Sandwich Microarray Assays

3.3.1. Assembling Slide Holder and Slide Preparation

1. Four preblocked slides are mounted in the slide holder.
2. The slides are washed once with 1X PBS/0.005% Tween-20 buffer.
3. Excess wash solution is removed by inverting the slide holder and spinning the slide holder briefly (285*g*, 30 s).

3.3.2. Sample Incubation

1. Dilutions of plasma samples and of standard mixtures for calibration curves are prepared in casein in a 384-well plate. For the most concentrated dilution, unprocessed plasma or freshly prepared mixtures of purified antigens are diluted 1:1 with 12.5 μL casein blocking solution. Subsequent dilutions are serially diluted 10-fold into casein using 2.4 μL of the first dilution.
2. Twenty microliters of plasma and standard samples are applied to each antibody array of the slide holder with a multi-channel pipettor or a laboratory robot (*see* **Note 17**).
3. After an incubation time of 1 h, all arrays are washed with 1X PBS/0.005% Tween-20 buffer.
4. Excess wash solution is removed by inverting the slide holder and spinning it briefly (285*g*, 30 s).
5. During all incubations, the solutions are mixed continuously by placing the slide holder on an orbital shaker (Tecan, Research Triangle Park, NC) in a climate-controlled room (1500 rpm, 20°C, 70% relative humidity) (*see* **Note 8**).

3.3.3. Incubation With Detection Antibodies

1. A mixture of biotinylated detection antibodies diluted into casein containing 0.5 mg/mL goat IgG is applied to the microarrays for 1 h with vigorous mixing followed by a wash step.
2. Excess wash solution is removed by centrifugation as described previously.

3.3.4. Staining With RLS Particles

1. Prior to staining with RLS particles, slides are re-blocked with casein solution for 10 min.
2. In a final 1-h incubation step, RLS gold particles are applied to the arrays at a final dilution of 0.25 OD_{600} in casein containing 300 m*M* NaCl and 0.5 mg/mL goat IgG.
3. Excess material is removed by washing.
4. The slide holder is disassembled, the slides are dipped twice into 50 mL deionized water, and are spun dry (400*g*, 1 min) in Falcon tubes before coating with RLS archiving solution (Invitrogen, Carlsbad, CA).

3.3.5. Archiving

1. To enhance the light-scattering properties of the RLS particles as well as to reduce light-scattering effects from surface scratches and dust particles, a polymer coating is applied to the slides.
2. Slides are coated with archiving solution by very slowly dipping each slide individually into a slide mailer three-quarters filled with the archiving solution (*see* **Note 19**).
3. Excess material is allowed to drip off the slide. Each slide is placed in a separate 50-mL Falcon tube, and excess archiving solution is removed by centrifugation (400*g*, 45 s) (*see* **Note 20**).

3.4. Microarray Imaging and Image and Data Analysis

1. Archived slides are typically scanned the next morning on a 16-bit CCD camera-based scanner optimized for the detection of RLS light scattering. Four slides at a time are loaded into the holder of the scanner. A resolution of 10 μm is chosen. The exposure time is adjusted with the standard curves so as to not saturate the 16-bit detector. Typical exposure times are 0.3 s.
2. After imaging, the TIFF file is loaded into ArrayVision. A mask with the array layout is applied to the image. The orientation and the dimensions of the mask are adjusted. Proper alignment of the mask with the spots of all subarrays is checked manually. Spots affected by artifacts such as dust and streaks are manually flagged and excluded from the analysis before signals are integrated. Raw median-trimmed mean (MTM) values (without subtraction of the local background) are exported to an Excel worksheet.
3. Raw MTM values are copied into the RAW data table in the analysis Excel file (**Fig. 1**). Within Excel, a Visual Basic program will convert the raw data, using

	A	B	C	D	E	F	G	H	I	J	K	L	M	N	O	P	Q	R
1	Spot labels	MTM Dens	% Removed	MAD - Lev	SD - Levels	MTM SD -	Pos X - mr	Pos Y - mı	Area - mm	Bkgd	sMTMDen	S/N	Flag	% At Floor	% At Ceili	% At Floor	% At Ceiling - Bkgd	
2	R1 - C1 : A : A - 1	50.23	13.043	11	59.92	14.74	2.744	18.322	0.007	95.422	0	0	0	0	0	0.025	0.023	
3	R1 - C1 : B : A - 1	473.03	10.784	168	416.75	191.99	2.943	18.308	0.011	95.422	377.611	0.372	0	0	0	0.025	0.023	
4	R1 - C1 : C : A - 1	1896.59	7.826	626	1222.47	770.96	3.146	18.31	0.012	95.422	1801.173	1.773	0	0	0	0.025		
5	R1 - C1 : D : A - 1	136.06	7.246	37	285.16	52.26	3.342	18.311	0.007	95.422	40.641	0.04	0	0	0	0.025		
6	R1 - C1 : E : A - 1	23737.42	0	4151	6160.85	6160.85	3.548	18.305	0.014	95.422	23642	23.273	0	0	0	0.025		
7	R1 - C1 : F : A - 1	10644.9	0	2689.5	3773.69	3773.69	3.74	18.305	0.01	95.422	10549.48	10.385	0	0	0	0.025		
8	R1 - C1 : G : A - 1	321.63	8.654	83	248.43	124.09	3.944	18.301	0.011	95.422	226.21	0.223	0	0	0	0.025		
9	R1 - C1 : H : A - 1	368.54	6.14	95	367.52	135.6	4.147	18.296	0.012	95.422	273.12	0.269	0	0	0	0.025		
10	R1 - C1 : I : A - 1	5770.4	2.381	1715	2679.87	2383.64	4.344	18.316	0.013	95.422	5674.977	5.586	0	0	0	0.025	0.023	
11	R1 - C2 : A : A - 1	51.55	13.043	11	33.24	15.7	7.242	18.301	0.007	95.422	0	0	0	0	0	0.025	0.023	
12	R1 - C2 : B : A - 1	327.31	11.304	87	423.22	118.48	7.462	18.308	0.012	95.422	231.892	0.228	1	0	0	0.025	0.023	
13	R1 - C2 : C : A - 1	2353.96	4.808	815	1340.91	1089.92	7.654	18.311	0.011	95.422	2258.538	2.223	0	0	0	0.025		
14	R1 - C2 : D : A - 1	182.29	5.797	58	152.2	82.67	7.851	18.301	0.007	95.422	86.871	0.086	0	0	0	0.025		
15	R1 - C2 : E : A - 1	29426.65	4.587	4033	6936.53	5872.5	8.062	18.309	0.012	95.422	29331.23	28.873	0	0	0	0.025		
16	R1 - C2 : F : A - 1	9071.66	0	2171.5	3072.59	3072.59	8.263	18.311	0.01	95.422	8976.235	8.836	0	0	0	0.025		
17	R1 - C2 : G : A - 1	339.69	9.677	90	255.96	118.74	8.463	18.299	0.01	95.422	244.269	0.24	0	0	0	0.025		
18	R1 - C2 : H : A - 1	141.97	7.246	43	115.1	64.32	8.665	18.291	0.007	95.422	46.547	0.046	0	0	0	0.025		
19	R1 - C2 : I : A - 1	2943.47	9.195	855	1869.14	1090.79	8.858	18.31	0.009	95.422	2848.047	2.804	0	0	0	0.025	0.023	
20	R1 - C3 : A : A - 1	57.45	10.145	13	50.21	17.14	11.74	18.28	0.007	95.422	0	0	0	0	0	0.025	0.023	
21	R1 - C3 : B : A - 1	163.84	15.942	50	151.97	60.9	11.946	18.28	0.007	95.422	68.423	0.067	0	0	0	0.025	0.023	
22	R1 - C3 : C : A - 1	2941.43	3.008	887	1422.46	1226.28	12.15	18.281	0.014	95.422	2846.005	2.802	0	0	0	0.025		
23	R1 - C3 : D : A - 1	290.6	10.714	99	502.29	124.28	12.336	18.238	0.003	95.422	195.178	0.192	0	0	0	0.025		
24	R1 - C3 : E : A - 1	23652.45	2.29	3612	6340.17	6031.86	12.555	18.277	0.014	95.422	23557.02	23.189	0	0	0	0.025		
25	R1 - C3 : F : A - 1	9708.86	1.02	1623.5	2914.9	2819.08	12.75	18.275	0.01	95.422	9613.434	9.463	0	0	0	0.025		
26	R1 - C3 : G : A - 1	248.16	7.547	53	163.83	70	12.934	18.265	0.006	95.422	152.742	0.15	0	0	0	0.025		
27	R1 - C3 : H : A - 1	566.94	23.81	190	705.41	277.07	13.162	18.315	0.002	95.422	471.516	0.464	0	0	0	0.025		
28	R1 - C3 : I : A - 1	5399.91	8	1747	3687.42	2505.21	13.352	18.285	0.013	95.422	5304.491	5.222	0	0	0	0.025	0.023	
29	R1 - C4 : A : A - 1	192.65	21.739	59	25828.05	65.22	16.258	18.27	0.007	95.422	97.227	0.096	1	0	18.841	0.025	0.023	
30	R1 - C4 : B : A - 1	352.23	17.391	130	24030.7	160.01	16.454	18.26	0.007	95.422	256.807	0.253	1	0	15.942	0.025	0.023	
31	R1 - C4 : C : A - 1	3597.81	0.962	907	1478.34	1426.63	16.666	18.271	0.011	95.422	3502.384	3.448	0	0	0	0.025		
32	R1 - C4 : D : A - 1	434.32	4.348	122	221.1	201.05	16.861	18.231	0.002	95.422	338.897	0.334	1	0	0	0.025		
33	R1 - C4 : E : A - 1	22179.36	0	4097	6426.88	6426.88	17.073	18.27	0.012	95.422	22083.94	21.739	0	0	0	0.025		
34	R1 - C4 : F : A - 1	10125.66	0	1972	3538.13	3538.13	17.269	18.271	0.01	95.422	10030.24	9.874	0	0	0	0.025		
35	R1 - C4 : G : A - 1	214.02	14.493	60	214.98	85.27	17.465	18.26	0.007	95.422	118.595	0.117	0	0	0	0.025		
36	R1 - C4 : H : A - 1	199.2	11.594	67	841.6	91.72	17.661	18.26	0.007	95.422	103.775	0.102	1	0	0	0.025		
37	R1 - C4 : I : A - 1	5598.3	5.208	1251.5	2329.24	1875.07	17.865	18.274	0.01	95.422	5502.875	5.417	0	0	0	0.025	0.023	
38	R1 - C5 : A : A - 1	77.67	15.942	13	171.13	19.11	20.756	18.249	0.007	95.422	0	0	0	0	0	0.025	0.023	
39	R1 - C5 : B : A - 1	419.89	11	104	345.61	144.3	20.961	18.258	0.011	95.422	324.466	0.319	0	0	0	0.025	0.023	
40	R1 - C5 : C : A - 1	3761.03	1.724	900	1565.52	1402.42	21.166	18.256	0.012	95.422	3665.605	3.608	0	0	0	0.025		
41	R1 - C5 : D : A - 1	190.11	11.594	51	205.84	60.56	21.355	18.249	0.007	95.422	94.693	0.093	0	0	0	0.025		
42	R1 - C5 : E : A - 1	31588.19	9.6	3517	7183.29	5197.66	21.575	18.249	0.013	95.422	31492.76	31.001	0	0	0	0.025		
43	R1 - C5 : F : A - 1	12411.57	0	2687.5	3903.15	3903.15	21.768	18.25	0.011	95.422	12316.15	12.124	0	0	0	0.025		
44	R1 - C5 : G : A - 1	230.76	15.942	52	225.02	66.21	21.963	18.239	0.007	95.422	135.337	0.133	0	0	0	0.025		
45	R1 - C5 : H : A - 1	650.22	2.703	191	488.13	287.11	22.196	18.25	0.004	95.422	554.801	0.546	1	0	0	0.025		
46	R1 - C5 : I : A - 1	5326.13	3.125	1285.5	2305.17	1927.82	22.361	18.255	0.01	95.422	5230.707	5.149	0	0	0	0.025	0.023	

Instructions / Setup / Raw / FittingData / All Samples / Summary Table / BestDilutions / Sensitivities / Manual

Ready NUM

Fig. 1. Excel data analysis macro. RAW data input from the image analysisprogram Array Vision.

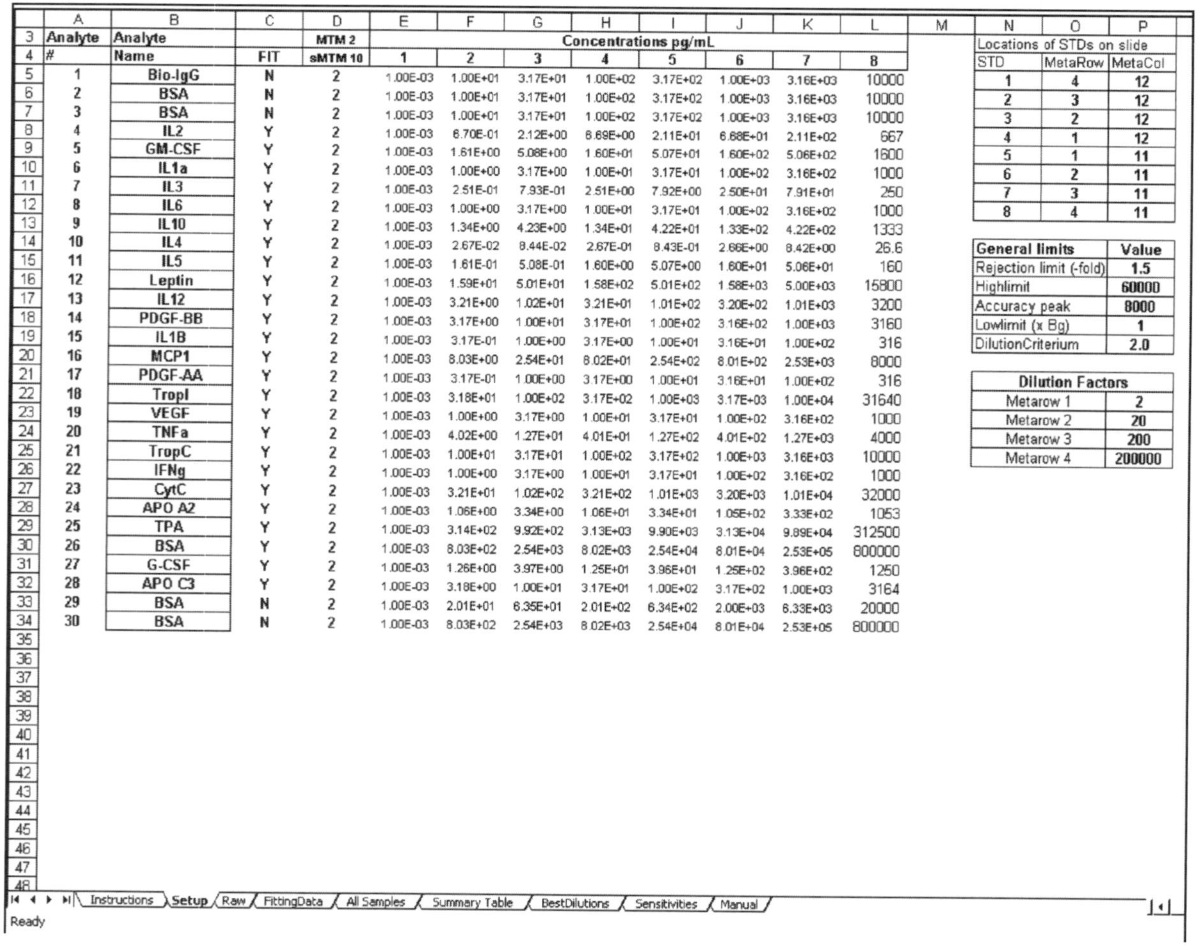

	A	B	C	D	E	F	G	H	I	J	K	L
3	Analyte	Analyte		MTM 2	Concentrations pg/mL							
4	#	Name	FIT	sMTM 10	1	2	3	4	5	6	7	8
5	1	Bio-IgG	N	2	1.00E-03	1.00E+01	3.17E+01	1.00E+02	3.17E+02	1.00E+03	3.16E+03	10000
6	2	BSA	N	2	1.00E-03	1.00E+01	3.17E+01	1.00E+02	3.17E+02	1.00E+03	3.16E+03	10000
7	3	BSA	N	2	1.00E-03	1.00E+01	3.17E+01	1.00E+02	3.17E+02	1.00E+03	3.16E+03	10000
8	4	IL2	Y	2	1.00E-03	6.70E-01	2.12E+00	6.69E+00	2.11E+01	6.68E+01	2.11E+02	667
9	5	GM-CSF	Y	2	1.00E-03	1.61E+00	5.08E+00	1.60E+01	5.07E+01	1.60E+02	5.06E+02	1600
10	6	IL1a	Y	2	1.00E-03	1.00E+00	3.17E+00	1.00E+01	3.17E+01	1.00E+02	3.16E+02	1000
11	7	IL3	Y	2	1.00E-03	2.51E-01	7.93E-01	2.51E+00	7.92E+00	2.50E+01	7.91E+01	250
12	8	IL6	Y	2	1.00E-03	1.00E+00	3.17E+00	1.00E+01	3.17E+01	1.00E+02	3.16E+02	1000
13	9	IL10	Y	2	1.00E-03	1.34E+00	4.23E+00	1.34E+01	4.22E+01	1.33E+02	4.22E+02	1333
14	10	IL4	Y	2	1.00E-03	2.67E-02	8.44E-02	2.67E-01	8.43E-01	2.66E+00	8.42E+00	26.6
15	11	IL5	Y	2	1.00E-03	1.61E-01	5.08E-01	1.60E+00	5.07E+00	1.60E+01	5.06E+01	160
16	12	Leptin	Y	2	1.00E-03	1.59E+01	5.01E+01	1.58E+02	5.01E+02	1.58E+03	5.00E+03	15800
17	13	IL12	Y	2	1.00E-03	3.21E+00	1.02E+01	3.21E+01	1.01E+02	3.20E+02	1.01E+03	3200
18	14	PDGF-BB	Y	2	1.00E-03	3.17E+00	1.00E+01	3.17E+01	1.00E+02	3.16E+02	1.00E+03	3160
19	15	IL1B	Y	2	1.00E-03	3.17E-01	1.00E+00	3.17E+00	1.00E+01	3.16E+01	1.00E+02	316
20	16	MCP1	Y	2	1.00E-03	8.03E+00	2.54E+01	8.02E+01	2.54E+02	8.01E+02	2.53E+03	8000
21	17	PDGF-AA	Y	2	1.00E-03	3.17E-01	1.00E+00	3.17E+00	1.00E+01	3.16E+01	1.00E+02	316
22	18	TropI	Y	2	1.00E-03	3.18E+01	1.00E+02	3.17E+02	1.00E+03	3.17E+03	1.00E+04	31640
23	19	VEGF	Y	2	1.00E-03	1.00E+00	3.17E+00	1.00E+01	3.17E+01	1.00E+02	3.16E+02	1000
24	20	TNFa	Y	2	1.00E-03	4.02E+00	1.27E+01	4.01E+01	1.27E+02	4.01E+02	1.27E+03	4000
25	21	TropC	Y	2	1.00E-03	1.00E+01	3.17E+01	1.00E+02	3.17E+02	1.00E+03	3.16E+03	10000
26	22	IFNg	Y	2	1.00E-03	1.00E+00	3.17E+00	1.00E+01	3.17E+01	1.00E+02	3.16E+02	1000
27	23	CytC	Y	2	1.00E-03	3.21E+01	1.02E+02	3.21E+02	1.01E+03	3.20E+03	1.01E+04	32000
28	24	APO A2	Y	2	1.00E-03	1.06E+00	3.34E+00	1.06E+01	3.34E+01	1.05E+02	3.33E+02	1053
29	25	TPA	Y	2	1.00E-03	3.14E+02	9.92E+02	3.13E+03	9.90E+03	3.13E+04	9.89E+04	312500
30	26	BSA	Y	2	1.00E-03	8.03E+02	2.54E+03	8.02E+03	2.54E+04	8.01E+04	2.53E+05	800000
31	27	G-CSF	Y	2	1.00E-03	1.26E+00	3.97E+00	1.25E+01	3.96E+01	1.25E+02	3.96E+02	1250
32	28	APO C3	Y	2	1.00E-03	3.18E+00	1.00E+01	3.17E+01	1.00E+02	3.17E+02	1.00E+03	3164
33	29	BSA	N	2	1.00E-03	2.01E+01	6.35E+01	2.01E+02	6.34E+02	2.00E+03	6.33E+03	20000
34	30	BSA	N	2	1.00E-03	8.03E+02	2.54E+03	8.02E+03	2.54E+04	8.01E+04	2.53E+05	800000

N	O	P
Locations of STDs on slide		
STD	MetaRow	MetaCol
1	4	12
2	3	12
3	2	12
4	1	12
5	1	11
6	2	11
7	3	11
8	4	11

General limits	Value
Rejection limit (-fold)	1.5
Highlimit	60000
Accuracy peak	8000
Lowlimit (x Bg)	1
DilutionCriterium	2.0

Dilution Factors	
Metarow 1	2
Metarow 2	20
Metarow 3	200
Metarow 4	200000

Instructions / Setup / Raw / FittingData / All Samples / Summary Table / BestDilutions / Sensitivities / Manual

Ready

Fig. 2. Excel data analysis macro. SETUP window defining analytes of a 24 mouse plasma antibody array, antigen concentrations (in pg/mL) used for generating calibration curves in eight microarrays as specified in the upper right table, as well as sample dilution factors and general settings.

inputs describing the array layout, concentrations of antigen standards, and dilution factors, into concentration readouts in pg/mL for each analyte and for each dilution.

4. Analyte names, location on the array, and the concentrations of antigens used for the calibration arrays, as well as the array layout and sample dilutions, are defined in the SETUP worksheet (**Fig. 2**).
5. The automatically fitted standard curves for each analyte are displayed in the MANUAL worksheet (**Fig. 3**). These eight-point calibration curves are inspected visually, and the fit can be modified by re-setting and restricting the fitting parameters and removing obvious outlier data points (*see* **Note 21**). Below the calibration curve, a table lists the raw signal and the concentration value for each of the four dilutions measured for each of the 10 samples that are being analyzed per antibody microarray slide (*see* **Note 22**). The program automatically chooses the lowest dilution that gives a signal in the dynamic range of the assay. Alternatively, a dilution can be selected manually, which often allows for a more accurate comparison across many samples, especially for assays where matrix effects result in dramatically different concentration readouts depending on which dilution is used.
6. After clicking on STORE DILUTION (**Fig. 3**), the results are exported to a SUMMARY data sheet (**Fig. 4**) that lists the concentration values for all analytes and samples of one microarray slide. For multi-slide experiments, these SUMMARY data sheets are combined and processed further.

3.5. Assay Development and Optimization

3.5.1. Reagent Selection and Characterization

1. To identify microarray-compatible sandwich pairs, several commercially available antibodies are obtained against each analyte of interest and tested in all possible combinations, with each antibody acting as capture and also as detection antibody (*see* **Note 23**).
2. All unmodified antibodies are arrayed out as capture antibodies and incubated with their individual biotinylated variants at 0.5 μg/mL in the absence or presence of increasing concentrations of single antigens (100 pg/mL, 10 ng/mL, 1 μg/mL) (*see* **Note 24**).
3. Antibodies associated with high antigen-independent background signals are discarded. The most sensitive sandwich pair remaining is selected for further testing and for the final panel assembly (*see* **Note 25**).

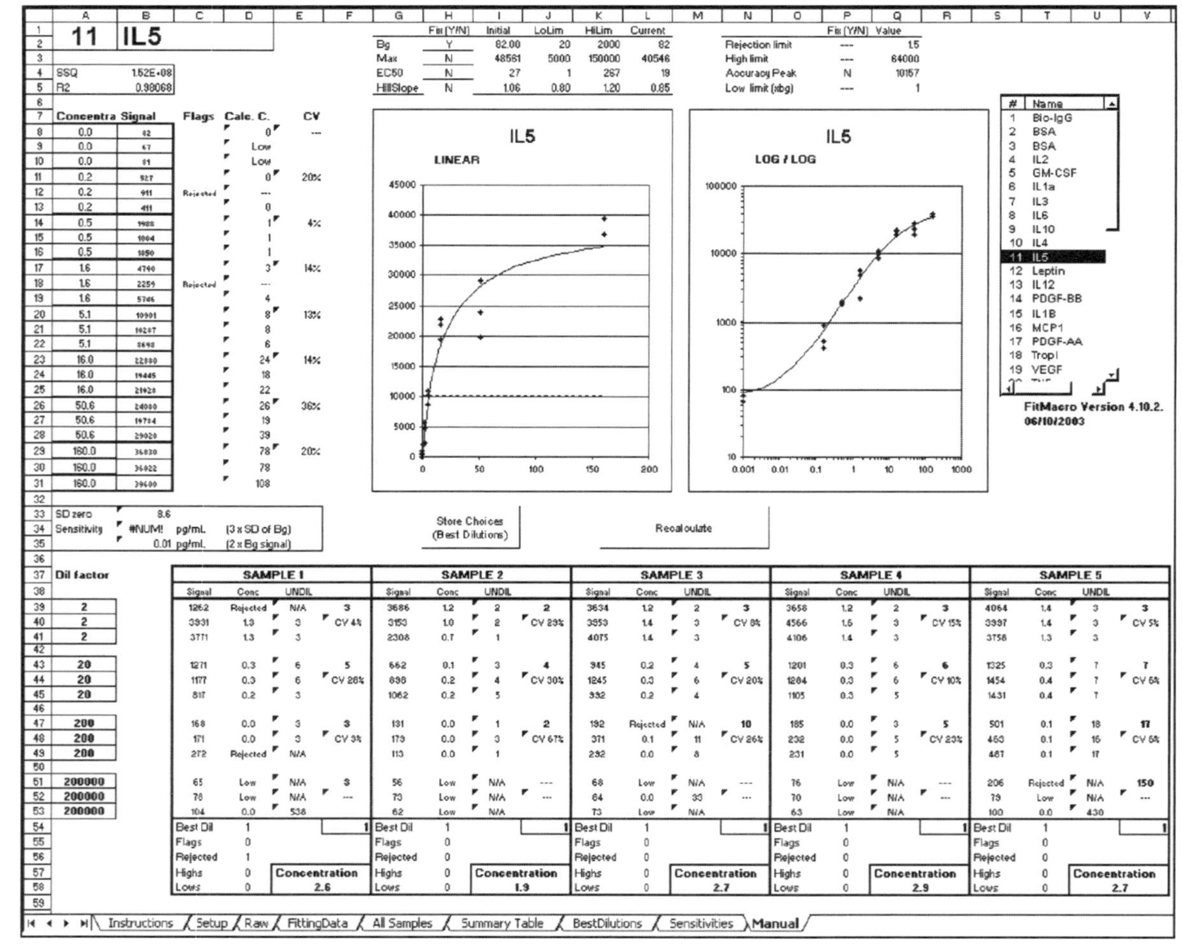
11 IL5
SSQ 1.52E+08
R2 0.98068
Bg
Max
EC50
HillSlope
Fit (Y/N) Initial LoLim HiLim Current
Rejection limit
High limit
Accuracy Peak
Low limit (xbg)
Concentra Signal
Flags
Calc. C.
CV
IL5
LINEAR
LOG / LOG
SD zero
Sensitivity
Store Choices (Best Dilutions)
Recalculate
Dil factor
SAMPLE 1
SAMPLE 2
SAMPLE 3
SAMPLE 4
SAMPLE 5
Signal Conc UNDIL
Best Dil
Flags
Rejected
Highs
Lows
Concentration
FitMacro Version 4.10.2.
06/10/2003
Instructions
Setup
Raw
FittingData
All Samples
Summary Table
BestDilutions
Sensitivities
Manual

Fig. 3. Excel data analysis macro. MANUAL window for mouse interleukin 5 (IL5) with the eight-point calibration curve displayed linear and log/log plot of resonance light scattering (RLS) signal vs analyte concentration in pg/mL. To the left of the curves, the raw data for each of the spots for each of the standard concentrations is shown. To the right of the curves, a scroll-down window allows the selection of different analytes. At the bottom, a table summarizes raw data and calculated concentrations in pg/mL for each of the four dilutions of 10 samples (for clarity, only five samples are shown).

Calculated Concentrations for All Analytes / Best Dilutions

080204_17

Analyte	Sample 1	Sample 2	Sample 3	Sample 4	Sample 5	Sample 6	Sample 7	Sample 8	Sample 9	Sample 10
Mouse ID	A-1	A-2	A-3	A-4	A-5	A-6	A-7	A-8	A-9	A-10
IL2	0.17	---	0.10	0.22	0.22	0.28	0.44	0.39	0.02	0.17
GM-CSF	9.49	5.28	4.13	0.61	3.77	1.76	5.70	---	2.22	0.35
IL1a	7.84	6.34	4.98	5.10	9.32	7.03	10.15	4.19	9.15	8.03
IL3	---	---	---	0.59	---	---	0.00	---	---	0.01
IL6	1.20	0.85	1.21	1.30	1.68	1.55	1.56	1.79	1.53	1.42
IL10	5.07	3.88	2.78	2.85	1.99	1.51	2.13	1.82	1.65	1.76
IL4	0.06	0.04	0.04	0.04	0.04	0.06	0.04	0.03	0.01	0.05
IL5	2.62	1.93	2.65	2.86	2.70	2.49	2.67	2.12	3.17	2.64
Leptin	233.73	27.11	179.17	190.64	36.71	272.88	297.55	50.25	289.68	143.90
IL12	928.41	721.25	598.67	685.07	1136.50	962.79	692.25	821.01	911.81	966.58
PDGF-BB	208.43	259.36	133.13	307.62	258.12	249.22	246.96	258.73	327.52	347.63
IL1B	0.55	0.07	0.15	0.11	0.29	0.30	0.15	0.04	0.12	0.07
MCP1	61.07	15.96	22.32	20.56	36.09	29.65	28.28	24.05	50.90	59.89
PDGF-AA	1696.76	967.50	1900.24	4025.54	1607.46	2497.35	1928.76	2757.07	4104.14	3147.98
TropI	---	---	---	---	---	---	---	---	---	---
VEGF	90.42	88.04	113.62	131.04	90.88	118.73	138.71	92.18	131.88	119.58
TNFa	20.90	10.16	63.90	84.12	59.11	53.96	108.05	12.99	251.71	79.22
TropC	69.15	126.71	189.82	269.86	124.95	87.59	206.37	115.33	98.48	88.42
IFNg	0.72	0.74	0.90	0.81	1.61	0.67	3.40	1.45	---	0.06
CytC	---	---	---	---	---	---	---	---	---	---
APO A2	---	---	770874.35	324227.96	10825.47	1124837.60	1091727.61	162710.08	2323649.72	949073.43
TPA	356850.86	2432381.05	618798.16	542656.08	2109381.10	1576697.68	1311737.30	382838.21	1038601.32	2306840.37
G-CSF	3.62	1.83	3.32	2.71	4.08	1.90	2.32	1.24	3.65	1.93
APO C3	644028.42	665747.92	210927.03	132397.57	855113.26	761631.05	656290.31	146437.16	620798.92	---

Instructions / Setup / Raw / FittingData / All Samples / **Summary Table** / BestDilutions / Sensitivities / Manual

Ready

Fig. 4. Excel data analysis macro. SUMMARY TABLE listing the concentrations (in pg/mL) of 24 mouse plasma proteins measured in 10 samples.

3.5.2. Iterative Panel Assembly and Optimization

1. For the assembly of multiplexed immunoassays, up to 45 capture antibodies (*see* **Note 1**) targeting distinct antigens are arrayed and incubated with a mixture of the 45 corresponding detection antibodies in the absence or presence of increasing concentrations of individual antigens (e.g., 100 pg/mL, 10 ng/mL, 1 µg/mL). This experimental setup will help evaluate the concentration-dependent cross-reactivity of each analyte with each individual sandwich assay on the array.
2. In iterative rounds of testing and reshuffling, individual immunoassays are assessed for their potential to work together in assemblies of 45. Generally cross-reactive and grossly incompatible immunoassays are discarded. The remaining immunoassays are grouped into separate sub-panels with a view to minimize antigen-dependent cross-reactivity. In this phase of assay optimization, generic antibody concentrations are used for spotting and detection.
3. To evaluate the antigen-independent antibody cross-talk in the final panel, each detection antibody is incubated individually at a concentration of 0.5 µg/mL in the absence of antigens with the entire capture antibody array.
4. To evaluate the antigen-dependent on-feature cross-reactivity in the final panel, the detection antibody mix is incubated with increasing concentrations of individual antigens.
5. Individual antigen titrations against the entire panel are used to adjust antigen concentrations to cover the dose-response curve of the immunoassay in an eight-point threefold dilution series.
6. Multiple standard curves are recorded with antigen and detection antibody mixtures, and the performance of individual immunoassays in the multiplexed format is compared to the assay performance using individual analytes with respect to background signal in the absence of antigen and with respect to the dynamic range of each assay. To reduce the effect of the increased antigen-independent background in the fully multiplexed assay, the concentrations of all detection antibodies are stepwise lowered as far as possible (*see* **Note 26**).
7. Finally, test measurements are performed with plasma samples to fine-tune the standard concentrations and panel composition as needed.
8. To a limited extent, the assay sensitivity can be enhanced by increasing the concentration of the capture antibody solution used in the printing process (*see* **Note 9**).

4. Notes

1. Although 48 antibody samples are printed in triplicates on 12 × 12 arrays, 3 samples are reserved for a biotinylated antibody to mark the array outline for easier image analysis.
2. Many alternative substrates are commercially available, but the spotting buffer must be adjusted to the hydrophobicity of the surface and to the temperature and the humidity of the printing environment in order to obtain good-quality spots. In the printing process, tight humidity control is essential to prevent concentrated antibody solutions from clogging the capillary pins as well as for good spot con-

sistency. The antibody solutions need to be free of particles, and near clean-room conditions are recommended. We compared the performance of multiplexed sandwich assays on a variety of commercially available slides (data not shown), including plain amino-silane-coated slides for noncovalent attachment, aldehyde- and amine-reactive slides for covalent attachment, as well as a number of slides coated in-house with an amine-reactive vinyl axazolone-polymer featuring different mixtures and types of copolymers ***(34)***. In terms of sensitivity of the assays, we saw little difference between different substrates; proper adjustment of the printing conditions to the hydrophobicity of the surface in order to obtain good spot morphology appeared in our hands more important than the choice of attachment chemistry. Physical adsorption to amino-silane-coated slides appears to perform sufficiently well for full-length antibodies.

3. The printer is situated in a temperature- and humidity-controlled room with HEPA-filtered air and "sticky" mats at the entrance to provide near clean-room conditions.
4. Despite several attempts, we have not been able to establish assays using antibody sera or unpurified antibodies as capture antibodies.
5. Care must be taken to not overly biotinylate the protein, as this may cause high background signal. Conditions are adjusted so that approximately two to four biotin molecules are conjugated to a single antibody molecule.
6. Depending on the applications, other matrices may be more advantageous.
7. The actual wash volume per well is 800 μL because the wash head was modified and features only 192 outlets. Shorter, gentler wash protocols should be tested for a given application.
8. Shaking the slide holder during incubation periods results in a more uniform background. To efficiently mix an assay volume of only 20 μL, the shaker must be operated at ≥1200 rpm.
9. Higher capture antibody concentrations will increase signal intensity ***(14)***, but nonspecific background and smearing of spots because of over-saturation of the surface may result. Capture antibodies can be spotted at concentrations as high as 1 mg/mL.
10. Print plates can be prepared up to 2 d prior to printing.
11. Depending on the print buffer and the hydrophobicity of the microarray substrate used, other environmental conditions may produce equally good results.
12. The effective temperature in the well is closer to 18°C. With the current settings, antibody solutions will not dry up or be diluted by condensation during print runs as long as 24 h.
13. SMP3 pins will deliver approx 180 spots using the current print buffer and conditions before the capillaries need to be refilled. For most consistent spot sizes, at least 50 spots should be preprinted and no more than 100 spots should be printed onto the microarray substrates. The spot diameter and shape will depend on the printing conditions, printing buffer, as well as the hydrophobicity of the microarray substrate. Using the reagents and protocols described, antibody spots are approx 120 μm in diameter.
14. SMP3 print tips are cleaned by dipping them into cleaning solution (VP110, V&P Scientific) followed by several rounds of drying with filtered nitrogen gas and

dipping into deionized water and finally isopropanol. Pins are inspected under a microscope for dirt and mechanical damage and wear. If necessary, the cleaning procedure is repeated several times or the pin is replaced.

15. If the slides are lowered too quickly, the samples will tend to smear across the slide.
16. Blocked antibody arrays can be stored dry in a desiccator at room temperature for more than 2 mo without loss in sensitivity and performance.
17. In all sample addition steps, care must be taken not to touch the bottom of the array. Solutions are added to the walls of the well and the slide holder is briefly spun to move the solution droplets to the bottom of the well and to remove air bubbles.
18. To reduce assay variations, wash solutions are completely removed from the slide holder by centrifugation after each wash step.
19. Extra care must be taken to avoid particulate matter such as dust from settling onto the slides. Contamination of the slide surface by particulate matter will cause streaks and light-scattering artifacts. The archiving solution needs to be applied as uniformly as possible.
20. Although slides are typically dried at room temperature overnight before scanning, they can also be scanned after quick drying in an oven. Fast-drying alternative archiving solutions are available from Invitrogen. Archiving also creates a permanent record of the assay, and slides can be re-scanned several times. Slides can be re-coated with archiving solution.
21. Calibration curves for each analyte are generated by fitting 24 data points (8 concentrations $\times$ 3 replicate spots) to a four-parameter logistic curve using $1/Y^2$ weighting. Unknown sample concentrations are calculated using the corresponding signal values, the curve parameters, and the dilution factors. The concentrations calculated for each of the three replicate spots on an array are compared to each other; if one of the spots gives a value 1.5-fold higher or lower than the median of the three, it is rejected. For the rest of the spots, an average is calculated. The rejection criterion was introduced in order to reduce the errors caused by randomly occurring spots of poor morphology, and it is applied to the measurement of samples as well as calibration standards.
22. Of the four dilutions, usually only one or two are within the dynamic range of the assay. To obtain a single concentration value, the program automatically chooses the lowest dilution that gives a raw signal at least twice the background.
23. For the development of our antibody arrays, over 350 antibodies and more than 120 protein standard preparations for over 100 antigens were tested.
24. These test arrays can contain sandwich pairs for several different analytes in parallel.
25. The successful antibody pairs can represent (a) combinations of monoclonal antibodies for antigen capture with monoclonal or polyclonal antibodies for detection, (b) polyclonal antibodies for capture with the identical polyclonal antibody for detection, or (c) soluble antigen receptor domains for capture with either monoclonal or polyclonal antibodies for detection. This early reagent selection phase screens for optimal assay reagents only on the level of the individual immunoassay. Individual detection antibodies are tested against a panel of cap-

ture antibodies targeting the identical antigen in the presence or absence of this same antigen. The cross-talk between individual immunoassays is not accounted for at this stage.

26. The antigen-independent on-feature background will increase with the number of detection antibodies in the mix. These nonspecific antibody–antibody interactions effectively limit the dynamic range as well as the sensitivity of the arrayed assays, and need to be minimized ***(14)***. This can be accomplished by establishing the lowest possible concentration of each detection antibody that maintains acceptable performance of each individual assay. The experiment to evaluate antigen-independent cross-talk of each detection antibody against the final capture antibody array is hereby used as a guide for which detection antibody is most critical in the optimization process. In the case that individual sandwich assays detect very abundant analytes or when very high-affinity antibodies are available, the detection antibody concentration, and in principle the capture antibody concentration as well, can be lowered substantially. Following this strategy, the sensitivity of the overall panel, and especially the sensitivity of problematic individual assays operating close to the detection limit, can be optimized in an iterative fashion at the cost of the sensitivity of other less critical assays. Detection antibody concentrations used in our panels range from 0.013 to 0.5 µg/mL.

Acknowledgments

B. H. G. would like to thank Ryan Okon, Dr. Joerg Eppinger, Dr. Masaki Warashina, and James Schmeits for their assistance in developing and optimizing protocols and methods presented here. Dr. David Carney was instrumental in the selection and testing of reagents for a mouse antibody microarray, and his help is greatly appreciated. Thank you to Matthew Schmid and Leanna Lagpacan for writing a Standard Operating Procedure that formed the basis of this chapter. Leanna Lagpacan continues to provide expert technical assistance. Special thanks go to Dr. Steve Roman and Dr. Todd Peterson of Invitrogen (formerly of Genicon Sciences), who closely collaborated with us in the early stages of the project and provided prototype reagents and instrumentation, and freely exchanged observations they had made in their internal developmental effort. The project was initiated by Dr. Peter Schultz (Institute Director, GNF) and B. H. G. would like to thank him and Dr. Scott Lesley (Director, GNF Protein Sciences) for continuing support.

References

1. Utz, P. J. (2005) Protein arrays for studying blood cells and their secreted products. *Immunological Reviews* **204,** 264–282.
2. Stoll, D., Templin, M. F., Bachmann, J., and Joos, T. O. (2005) Protein microarrays: Applications and future challenges. *Curr. Opin. Drug Disc. Devel.* **8,** 239–252.

3. Haab, B. B. (2001) Advances in protein microarray technology for protein expression and interaction profiling. *Curr. Opin. Drug Disc. Devel.* **4,** 116–123.
4. MacBeath, G. (2002) Protein microarrays and proteomics. *Nat. Genet.* **32,** 526–532.
5. Paweletz, C. P., Charboneau, L., Bichsel, V. E., Simone, N. L., Chen, T., Gillespie, J. W., et al. (2001) Reverse phase protein microarrays which capture disease progression show activation of pro-survival pathways at the cancer invasion front. *Oncogene* **20,** 1981–1989.
6. Nishizuka, S., Charboneau, L., Young, L., Major, S., Reinhold, W. C., Waltham, M., et al. (2003) Proteomic profiling of the NCI-60 cancer cell lines using new high-density reverse-phase lysate microarrays. *Proc. Natl. Acad. Sci. USA* **100,** 14,229–14,234.
7. Kononen, J., Bubendorf, L., Kallioniemi, A., Barlund, M., Schraml, P., Leighton, S., et al. (1998) Tissue microarrays for high-throughput molecular profiling of tumor specimens. *Nat. Med.* **4,** 844–847.
8. Chan, S. M., Ermann, J., Su, L., Fathman, C. G., and Utz, P. J. (2004) Protein microarrays for multiplex analysis of signal transduction pathways. *Nat. Med.* **10,** 1390–1396.
9. Joos, T. O., Schrenk, M., Hopfl, P., Kroger, K., Chowdhury, U., Stoll, D., et al. (2000) A microarray enzyme-linked immunosorbent assay for autoimmune diagnostics. *Electrophoresis* **21,** 2641–2650.
10. Robinson, W. H., DiGennaro, C., Hueber, W., Haab, B. B., Genovese, M. C., Muller, S., et al. (2001) Antigen microarray characterization of the autoantibody response in systemic lupus erythematosus and related diseases. *Arthritis Rheum.* **44,** S399.
11. Robinson, W. H., DiGennaro, C., Hueber, W., Haab, B. B., Kamachi, M., Dean, E. J., et al. (2002) Autoantigen microarrays for multiplex characterization of autoantibody responses. *Nat. Med.* **8,** 295–301.
12. Miller, J. C., Zhou, H., Kwekel, J., Cavallo, R., Burke, J., Butler, E. B., et al. (2003) Antibody microarray profiling of human prostate cancer sera: Antibody screening and identification of potential biomarkers. *Proteomics* **3,** 56–63.
13. Joos, T. O., Stoll, D., and Templin, M. F. (2002) Miniaturised multiplexed immunoassays. *Curr. Opin. Chem. Biol.* **6,** 76–80.
14. Nielsen, U. B. and Geierstanger, B. H. (2004) Multiplexed sandwich assays in microarray format—Review. *J. Immunol. Meth.* **290,** 107–120.
15. Huang, J. X., Mehrens, D., Wiese, R., Lee, S., Tam, S. W., Daniel, S., et al. (2001) High-throughput genomic and proteomic analysis using microarray technology. *Clin. Chem.* **47,** 1912–1916.
16. Mendoza, L. G., McQuary, P., Mongan, A., Gangadharan, R., Brignac, S., and Eggers, M. (1999) High-throughput microarray-based enzyme-linked immunosorbent assay (ELISA). *BioTechniques* **27,** 778.
17. Moody, M. D., Van Arsdell, S.W., Murphy, K.P., Orencole, S.F., and Burns, C. (2001) Array-based ELISAs for high-throughput analysis of human cytokines. *BioTechniques* **31,** 186–194.

18. Nielsen, U. B., Cardone, M. H., Sinskey, A. J., MacBeath, G., and Sorger, P. K. (2003) Profiling receptor tyrosine kinase activation by using Ab microarrays. *Proc. Natl. Acad. Sci. USA* **1000,** 9330–9335.
19. Pawlak, M., Schick, E., Bopp, M. A., Schneider, M. J., Oroszlan, P., and Ehrat, M. (2002) Zeptosens' protein microarrays: a novel high performance microarray platform for low abundance protein analysis. *Proteomics* **2,** 383–393.
20. Schweitzer, B., Roberts, S., Grimwade, B., Shao, W., Wang, M., Fu, Q., et al. (2002) Multiplexed protein profiling on microarrays by rolling-circle amplification. *Nat. Biotech.* **20,** 359–365.
21. Tam, S. W., Wiese, R., Lee, S., Gilmore, J., and Kumble, K. D. (2002) Simultaneous analysis of eight human Th1/Th2 cytokines using microarrays. *J. Immunol. Meth.* **261,** 157–165.
22. Wiese, R., Belosludtsev, Y., Powdrill, T., Thompson, P., and Hogan, M. (2001) Simultaneous multianalyte ELISA performed on a microarray platform. *Clin. Chem.* **47,** 1451–1457.
23. Wang, C. C., Huang, R.-P., Sommer, M., Lisoukov, H., Huang, R., Lin, Y., Miller, T., et al. (2002) Array-based multiplexed screening and quantitation of human cytokines and chemokines. *J. Proteome Res.* **1,** 337–343.
24. Ekins, R. P. (1998) Ligand assays: from electrophoresis to miniaturized microarrays. *Clin. Chem.* **44,** 2015–2030.
25. Saviranta, P., Ryan, O., Brinker, A., Warashina, M., Eppinger, J., and Geierstanger, B. H. (2004) Evaluating sandwich immunoassays in microarray format in terms of the ambient analyte regime. *Clin. Chem.* **50,** 1907–1920.
26. Ekins, R. P., Chu, F. W., and Biggart, E. (1989) Development of microspot multianalyte ratiometric immunoassay using dual fluorescent-labelled antibodies. *Anal. Chim. Acta* **227,** 73–96.
27. Yguerabide, J. and Yguerabide, E. E. (1998) Light-scattering submicroscopic particles as highly fluorescent analogs and their use as tracer labels in clinical and biological applications. I. Theory. *Anal. Biochem.* **262,** 137–156.
28. Yguerabide, J. and Yguerabide, E. E. (1998) Light-scattering submicroscopic particles as highly fluorescent analogs and their use as tracer labels in clinical and biological applications. II. Experimental characterization. *Anal. Biochem.* **262,** 157–176.
29. Yguerabide, J. and Yguerabide, E. E. (2001) Resonance light scattering particles as ultrasensitive labels for detection of analytes in a wide range of applications. *J. Cell. Biochem.* **Suppl. 37,** 71–81.
30. Bao, P., Frutos, A. G., Greef, C., Lahiri, J., Muller, U., Peterson, T. C., et al. (2002) High-sensitivity detection of dna hybridization on microarrays using resonance light scattering. *Anal. Chem.* **74,** 1792–1797.
31. Nam, J.-M., Thaxton, C. S., and Mirkin, C. A. (2003) Nanoparticle-based bio-bar codes for ultrasensitive detection of proteins. *Science* **301,** 1884–1886.
32. Winssinger, N., Damoiseaux, R., Tully, D. C., Geierstanger, B. H., Burdick, K., and Harris, J. L. (2004) PNA-encoded protease substrate microarrays. *Chem. Biol.* **11,** 1351–1360.

33. Haab, B. B., Geierstanger, B. H., Michailidis, G., Vitzthum, F., Forrester, S., Okon, R., et al. (2005) Immunoassay and antibody microarray analysis of the HUPO PPP reference specimens: systematic variation between sample types and calibration of mass spectrometry data. *Proteomics* **5,** 3278–3291.
34. Tully, D. C., Roberts, M. J., Geierstanger, B. H., and Grubbs, R. B. (2003) Synthesis of reactive poly(vinyl axazolones) via nitroxide-mediated "living" free radical polymerization. *Macromolecules* **36,** 4302–4308.

4

Chemical Proteomics Profiling of Proteasome Activity

Martijn Verdoes, Celia R. Berkers, Bogdan I. Florea, Paul F. van Swieten, Herman S. Overkleeft, and Huib Ovaa

Summary

Proteolysis is a key mechanism for protein homeostasis in living cells. This process is effected by different classes of proteases. The proteasome is one of the most abundant and versatile proteases, bearing three different proteolytic active sites. The proteasome plays an important role in essential biological pathways such as antigen presentation, signal transduction, and cell-cycle control feedback loops. The aim of this work is to design novel chemical strategies for capturing, detection, identification, and quantification—in one word, profiling—the active protease fractions of interest, in cells of different phenotypes. Here, a set of chemistry-based functional proteomics techniques is demonstrated by profiling the multi-catalytic protease activities of the proteasome. Importantly, functional profiling is complementary to expression level profiling and is an indispensable parameter for better understanding of mechanisms underlying biological processes.

Key Words: Functional profiling; proteomics; proteasome, proteasome inhibitors; Staudinger ligation; EL4 cells.

1. Introduction

Proteolysis, or the processing and degradation of proteins, has emerged as one of the most widely studied processes in biology today. Long viewed as a dead-end process, of importance only for the removal of obsolete peptides and proteins, proteolytic events are now associated with numerous biological events apart from the processing and degradation of polypeptides, ranging from protein maturation, signal transduction, cell-cycle control, and controlled cell death, to the generation and presentation of antigenic peptides in the adaptive immune system. Realizing that proteolysis plays an important role in many biochemical processes, it has become apparent that proteolytic events, and the enzymatic activities involved, are tightly controlled. This control can be exerted at the transcriptional level, but also on the maturation state and localization of

From: *Methods in Molecular Biology, Vol. 328: New and Emerging Proteomic Techniques*
Edited by: D. Nedelkov and R. W. Nelson © Humana Press Inc., Totowa, NJ

proteolytic activities. A case in point is the caspase family of proteolytic activities. Normally dormant, these cysteine proteases are triggered by a variety of signals and, upon activation, are involved in inducing apoptosis through a series of tightly controlled pathways.

The processing and presentation of antigenic polypeptides can be divided into two distinct pathways (for a comprehensive review on both pathways, *see* **refs. *1–5***). In the class I pathway, endogenous proteins are targeted for proteasomal degradation through polyubiquitination. Upon docking of the poly-ubiquitinated proteasome substrates to the 19S cap, the protein is unfolded and transferred into the inner 20S core, which harbors the six individual catalytic subunits, two copies of each of three distinct activities named hereafter as β1, β2, and β5. These endopeptidase activities residing in the 20S core cleave polypeptide substrates to produce oligopeptide fragments of typically 3 to 20 amino acid residues in length ***(6)***. Further processing and degradation is effectuated by a series of aminopeptidase activities, which act upon the N-terminal site of the proteasome products and cleave off a few amino acid residues. In this way, the vast majority of cytosolic and nuclear proteins are processed and degraded completely to produce amino acids for ensuing incorporation in newly synthesized proteins.

A small number of proteasome products, or their aminopeptidase-trimmed counterparts, escape to the endoplasmic reticulum with the help of the transporter associated with antigen presentation (TAP). Within the endoplasmic reticulum (ER) lumen, these may be further processed by ER-specific aminopeptidases to a final length of eight or nine amino acid residues and are finally loaded onto major histocompatibility complex (MHC) class I complexes that are assembled in the ER. Presentation of the loaded MHC class I peptide complexes allows the immune system to sample the nature of the cytosolic and nuclear protein content, which may contain exogenous proteins such as viral proteins. Of note is the presence in immunocompetent cells of a second proteasome particle, dubbed the immunoproteasome (as opposed to the constitutive proteasome expressed uniformly throughout the organism), which closely resembles the constitutive proteasome but now contains three distinct proteolytic activities named β1i, β2i, and β5i ***(6–9)***.

The class II pathway is responsible for the presentation to the immune system of extracellular proteins. Professional antigen-presenting cells (APCs) are charged with this task and do so by means of endocytic uptake of extracellular material and ensuing stepwise endolysosomal proteolysis through numerous proteolytic activities such as cysteine and aspartic acid proteases of the cathepsin family and asparagine endopeptidase (AEP). In a fashion rather resembling the class I pathway, MHC class II protein complexes are loaded with oligopeptides, ranging from 8 to 13 residues in length, resulting from the action of above-mentioned proteases. Fully assembled MHC II-peptide complexes are displayed on the outer cell membrane, enabling the immune system to monitor their contents. A further similarity between the two pathways is the control exerted on the proteolytic activities. In the class I pathway, the proteasomal catalytic activities are shielded from the intracellular protein content thanks to the barrel-shaped architecture of the 20S core. Protein substrates can reach the catalytic sites only through a series of modification, recognition, and unfolding events. Similarly, class II protein substrates are brought into contact with respective proteolytic activities by means of active sampling, by APCs, of their extracellular surrounding. Proteolytic control is exerted by the stepwise fusion of endocytic compartments with vesicles containing proteases. Many of these proteases are synthesized in a proform and are activated upon removal of an oligopeptide stretch by either autocatalytic cleavage or by the action of another activity, and many of the endosomal and lysosomal proteolytic activities display a pH optimum. Of interest are the recent reports describing the existence of cross-talk between the class I and class II pathways. Although originally thought of as separate routes dedicated to the presentation of either endogenic or exogenic antigens, it is now clear that oligopeptides derived from extracellular origin can end up being displayed by MHC class I assemblies. The exact nature of the mechanism of this process, however, remains to be established: what events trigger the occurrence of this "cross-presentation," and which proteolytic activities and cellular mechanisms are involved in the generation of exogenous class I antigenic peptides?

Cross-presentation is a relatively unexplored topic in the field of antigenic peptide presentation, and little is known about the role of proteolytic activities. The full extent to which proteolytic action is involved in cross-presentation remains to be established. Do additional aminopeptidases active in the class I pathway exist, both of cytosolic and ER origin, and what is the role of the individual cathepsins in the class II processing and presentation? Without much doubt, other yet-to-be-discovered proteolytic entities are involved. These questions and the current state of the art in addressing these are amply reviewed elsewhere and are not the subject of this chapter. Here we discuss some recently emerged sets of proteomics techniques that provide ways to study and identify proteolytic activities with similar properties in one sweep, and based on their activity, rather than on their expression levels. Globally, these strategies are based on the design of broad-spectrum, covalent, and irreversible inhibitors of the targeted proteolytic activities in question ***(10–12)***. These inhibitors, normally derived from oligopeptides and endowed with an electrophilic trap replacing the amide bond normally targeted by the proteases, can be tracked by means of an affinity tag. Through this tag, which can be either incorporated directly into the peptide-based probes (biotin, fluorescent tags) or attached at a later stage by means of a bio-orthogonal organic transformation, allows the study of targeted proteolytic activities by analytical techniques (gel electrophoresis, mass spectrometry; *see* **Fig. 1**).

In recent years, such techniques have become available through the endeavors of various research groups ***(10–16)***. Under **Subheadings 2.** and **3.**, the various methodologies are discussed in detail. Attention will be focused both on synthetic aspects (i.e., the preparation of the following probes: AdaK(Bio)Ahx_3L_3VS (Fig. 1A; *see also* ref. ***17***), Ada[125]IYAhx_3L_3VS (**Fig. 1B;** *see also* **ref. *17***), AdaAhx(α-N_3)Ahx_2L_3VS (**Fig. 1C**; *see also* **ref. *12***), and DansylAhx_3L_3VS (**Fig. 1D**; *see also* **ref. *18***) and application of these tools devised to study proteasome activity (*see* **Subheadings 2.10–2.13.**). The labeling results are obtained with EL4 or HeLa cells, but the methods are in principle applicable to the cell line of choice.

2. Materials

2.1. General Materials

1. Amino acid building blocks and peptide synthesis reagents (piperidine, diisopropylethylamine, hydroxybenzotriazole [HOBt], trifluoroacetic acid [TFA]) (Advanced ChemTech, Peptides International, Novabiochem, and American Bioanalytical).
2. Adamantane-acetic acid, triisopropylsilane [TIS], and iodoacetonitrile (Aldrich; *see* **Note 1**).
3. 4-Sulfamylbutyrylamimomethyl polystyrene resin (Novabiochem AG).
4. Iodo-gen (Pierce).
5. High-performance liquid chromatography (HPLC)-grade organic solvents dimethyl formamide [DMF], dichloromethane [DCM], acetonitrile [ACN], *N*-methylpyrrolidinone [NMP] (American Bioanalytical), and ethyl acetate [EtOAc] (Fisher) were used as received.
6. Sep-pak C_{18} columns (Waters).
7. Unless indicated otherwise, chemicals were obtained commercially of the highest available grade.

2.2. Cell Culture and Lysis

1. EL4 (mouse thymoma), HeLa (human cervix epitheloid carcinoma), and U373 (human astrocytoma) cells lines were used.
2. RPMI 1640 medium (Gibco, Invitrogen Corp.), Dulbecco's modified Eagle's medium (DMEM), L-glutamine, 10% fetal calf serum (FCS), penicillin/streptomycin solution, phosphate-buffered saline (PBS), and trypsin-ethylenediamine tetraacetic acid (EDTA) solution.
3. Lysis buffer for protease active cell lysates: 250 m*M* sucrose, 50 m*M* Tris-HCl pH 7.5, 5 m*M* $MgCl_2$, 1 m*M* dithiothreitol [DTT], 2 m*M* ATP in water.
4. Glass beads (140 μm).
5. NP-40 lysis buffer: 50 m*M* Tris-HCl (pH 8.0), 150 m*M* NaCl, 1% NP-40 was used to lyse the inhibitor-treated living cells used in **Subheadings 2.10–2.13.**

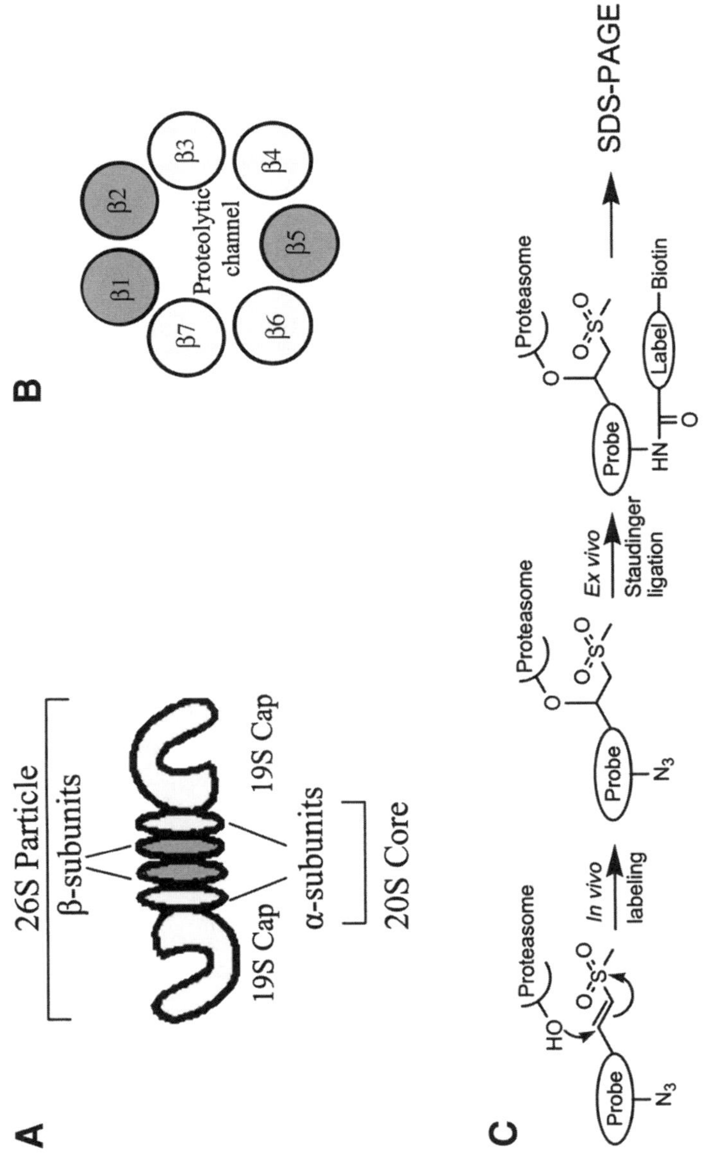

Fig. 1. **(A,B)** Schematic representation of the proteasome. **(C)** Two-step labeling of the proteasome by means of the Staudinger ligation.

2.3. Two-Dimensional Electrophoresis

1. Rehydration solution: 8 *M* urea, 2% CHAPS, 0.5% immobilized pH gradient (IPG) buffer (Amersham Pharmacia Biotech) (pH 3–10), bromophenol blue, 18 m*M* DTT.
2. For two-dimensional, nonequilibrium pH gradient sodium dodecyl sulfate-polyacrylamide get electrophoresis (2D NEPHGE SDS-PAGE): add 220 μL of rehydration solution to 30 μL of samples and mix.
3. Subject to first-dimension isoelectric focusing using a IPGphor isoelectric focusing system (Amersham Pharmacia Biotech) and second-dimension SDS-PAGE (12.5%). (The results obtained were comparable with conventional 2D NEPHGE SDS-PAGE methods.)

2.4. SDS-PAGE Separation and Western Blotting

1. SDS-PAGE gels of 12.5% polyacrylamide running gel and 4% stacking gel (BioRad).
2. Polyvinylidene difluoride (PVDF) membranes (Hybond P, Amersham).
3. Tris-buffered saline with Tween (TBS-T): prepare 10X stock with 1.37 *M* NaCl, 27 m*M* KCl, 250 m*M* Tris-HCl (pH 7.4), 1% Tween-20. Dilute 100 mL with 900 mL water for use.
4. Anti-dansyl-sulfonamidohexanoyl polyclonal antibody in TBS-T supplemented with 0.5% bovine serum albumin (BSA).
5. Horseradish peroxidase (HRP)-coupled goat anti-rabbit secondary antibody (Southern Biotech) in TBS-T supplemented with 0.5% BSA.
6. Streptavidin-HRP conjugate (Molecular Probes) in TBS-T (0.1% [v/v] Tween-20; *see* **Note 2**).
7. Enhanced chemiluminescence kit (ECL+, Amersham).

2.5. Synthesis of Probe A: AdaK(Bio)Ahx_3L_3VS (A)

1. Fmoc-Leu-Wang resin (0.5 mmol), Fmoc-Leu-OH, Fmoc-Ahx-OH, Fmoc-Ahx-OH, Fmoc-Ahx-Oh, Fmoc-Ahx-Oh, Fmoc-Lys(biotin)-OH, and adamantaneacetic acid.
2. PyBOP (activating agent) and diisopropylethyl amine (base).
3. Piperidine (20%) in dimethylformamide (DMF), TFA/TIS/H_2O 95/2.5/2.5 v/v/v.
4. Leucine vinyl sulfone TFA salt, HBTU, and diisopropylethylamine (DiPEA) in DMF.
5. Ethyl acetate, diethyl ether, hexanes, 10% to 20% methanol in methylene chloride.

2.6. Synthesis and Radiolabeling of Probe B: AdaY(^{125}I)Ahx_3L_3VS (B)

1. Fmoc-Tyr(OtBu)-OH instead of Fmoc-Lys(biotin)-OH.
2. Iodo-gen (100 μg; *see* **Note 3**) was dissolved in phosphate buffer (50 m*M*, pH 7.4, 10 μL).
3. AdaYAhx_3L_3VS (20 μg) was dissolved in acetonitrile (30 μL).
4. Na^{125}I (10 μL aqueous solution, 1 mCi), Sep-pak C_{18} column, and acetonitrile.

2.7. Synthesis of Probe C: AdaAhx(α-N_3)Ahx_2L_3VS (C)

1. *N*-(α-Boc),*N*-(ε-Fmoc)-l-lysine (2.0 g, 4.2 mmol).
2. 50% v/v TFA/CH_2Cl_2, trifluoromethanesulfonyl azide, 1 *M* HCl, EtOAc.

3. Wang resin (1.0 g, 0.86 mmol), Fmoc-Leu-OH (1.2 g, 3.4 mmol) in CH_2Cl_2 (25 mL), diisopropylcarbodiimide (0.54 mL, 3.44 mmol), a catalytic amount of 4-(dimethylamino)pyridine, CH_2Cl_2-MeOH, CH_2Cl_2, and Et_2O.
4. Resin-bound AdaAhx(α-N_3)Ahx_2L_2, TFA/H_2O 95/5 v/v, leucine vinyl sulfone TFA salt (1 eq), HBTU (1 eq), DiPEA (2.2 eq) in DMF, EtOAc.

2.8. Synthesis of the Staudinger Ligation Handle

1. Fmoc Rink amide resin (128 mg, 100 μmol), *N*-(α-Boc),*N*-(ε-Mtt)-l-lysine (0.5 mmol, 312 mg) was coupled.
2. 1% TFA in CH_2Cl_2 for 30 s, 10% DiPEA in DMF.
3. Biotin (0.5 mmol, 122 mg), PyBOP (1 eq), DiPEA (1.2 eq) in DMF.
4. Me_3P (0.6 mL of a 1 *M* solution in toluene, 0.6 mmol, 6 eq) in dioxane/water (4/1 v/v, 2 mL), dioxane, Fmoc-Ahx-OH (0.5 mmol, 180 mg).
5. Phosphine **2**, **Fig. 4** (90 mg, 0.25 mmol), EDC (48 mg, 0.25 mmol), hydroxybenzotriazole (HOBt) (41 mg, 0.3 mmol) in CH_2Cl_2 (2 mL)

2.9. Synthesis of Probe D: DansylAhx$_3$L$_3$VS (D)

Wang resin, *N*-(3-dimethylaminopropyl)-*N′*-ethylcarbodiimide hydrochloride, (S,*E*)-5-methyl-1-(methylsulfonyl)hex-1-en-3-amine (leucine vinyl sulfone, LeuVS).

2.10. Proteasome Profiling Using AdaK(Bio)Ahx$_3$L$_3$VS (A)

1. AdaK(Bio)Ahx_3L_3VS (A).
2. Cell lysate.
3. 3X SDS sample buffer: 0.67 mL 20% SDS, 1 mL 0.625 *M* Tris-HCl (pH 6.8), 2.1 mL 87% glycerol, 0.4 mL β-mercaptoethanol, 0.1 mL 10% phenol blue (*N,N*-dimethylindoaniline), 0.73 mL water.
4. Rehydration solution: 8 *M* urea, 2% CHAPS, 0.5% IPG buffer (Amersham Pharmacia Biotech) (pH 3.0–10.0), bromophenol blue, 18 m*M* DTT.

2.11. Proteasome Profiling Using AdaY(^{125}I)Ahx$_3$L$_3$VS (B)

1. AdaY(^{125}I)Ahx_3L_3VS (B).
2. Cell lysate.
3. 3X SDS sample buffer: 0.67 mL 20% SDS, 1 mL 0.625 *M* Tris-HCl (pH 6.8), 2.1 mL 87% glycerol, 0.4 mL β-mercapto-ethanol, 0.1 mL 10% bromophenol blue, 0.73 mL water.
4. Rehydration solution: 8 *M* urea, 2% CHAPS, 0.5% IPG buffer (Amersham Pharmacia Biotech) (pH 3.0–10.0), bromophenol blue, 18 m*M* DTT.

2.12. Proteasome Profiling in Living Cells Using Two-Step Labeling With AdaAhx(α-N$_3$)Ahx$_2$L$_3$VS (C)

1. Cells in culture.
2. AdaAhx(α-N_3)Ahx_2L_3VS (C)
3. Glass beads.

4. Lysis buffer: 50 m*M* Tris, 5 m*M* $MgCl_2$, 0.5 m*M* EDTA, 0.25 m*M* sucrose (pH 7.4).
5. Staudinger ligation handle.
6. 20% SDS solution.
7. 4X sample buffer: 1 mL 20% SDS, 1 mL 0.625 *M* Tris-HCl (pH 6.8), 2.1 mL 87% glycerol, 0.4 mL β-mercaptoethanol, 0.1 mL 10% bromophenol blue, 0.4 mL water.

2.13. Proteasome Profiling in Living Cells Using DansylAhx$_3$L$_3$VS (D)

1. Cells in culture.
2. DansylAhx$_3$L$_3$VS (D).
3. Glass beads.
4. NP-40 lysis buffer: 50 m*M* Tris-HCl (pH 8.0), 150 m*M* NaCl, 1% NP-40.

3. Methods

In this section, we describe several methods for the synthesis of selective proteasome inhibitors and their use in functional proteomics methodologies. In general, the protease inhibitors were dissolved in dimethylsulfoxide (DMSO) to 20 m*M* stock solutions that were further diluted in the buffer of interest. The active protease lysates should be kept on ice at all times to avoid protein degradation. Lysates were frozen as 50-μL aliquots at –80°C and used only once, preventing repeated freeze/thawing cycles. In some cases, high concentration of chromatin, DNA, or RNA might influence the outcome of the experiments. Then, post-nuclear lysate procedures should be explored either by using the NP-40 buffer or other mild methods like dounce homogenization, which selectively disrupts the cell membrane, leaving nuclei and other organelles intact.

3.1. General Methods

1. Leucine vinyl sulfone was prepared as reported ***(13,19)***.
2. Solid-phase peptide synthesis was carried out using a 180° Variable Rate Shaker (Peptides International).
3. Nuclear magnetic resonance (NMR) spectra were recorded on Varian (200 MHz, 500 MHz) and Bruker (200 MHz, 300 MHz, 400 MHz) spectrometers.
4. Mass spectra were recorded on an electrospray LCT liquid chromatography (LC)-mass spectrometry (MS) instrument (Waters).
5. Preparative reverse-phase HPLC purifications were carried out using a Waters PrepLC™ C18 column (250 mm × 40 mm) with a solvent gradient ranging from 30 to 95% acetonitrile in water containing 0.1% formic acid.

3.2. Cell Culture and Lysis

1. The cell lines EL4 (mouse thymoma) was cultured in RPMI 1640 medium (Gibco, Invitrogen Corp.).
2. HeLa cells (human cervix epitheloid carcinoma) and U373 cells (human astrocytoma) were cultured in DMEM.

3. Both media were supplemented with L-glutamine, 10% FCS, penicillin, and streptomycin.
4. Cells were grown at 37°C and 5% CO_2 in a humidified incubator.
5. Cells were harvested and lysed by glass beads or sonication in lysis buffer or by incubation on ice for 30 min in NP-40 lysis buffer followed by 5 min centrifugation at 16,000*g* to remove membrane fractions, nuclei and cell debris.

3.3. SDS-PAGE Separation and Western Blotting

1. Electro-transfer protein from SDS-PAGE gel onto PVDF membranes (Hybond P, Amersham).
2. Dansyl: anti-dansyl-hexanoylsulfonamido immunoblotting was performed using a rabbit anti-dansyl-sulfonamidohexanoyl polyclonal antibody (1:7500, 1 h at room temperature; *see* **Note 4**) and HRP–coupled goat anti-rabbit secondary antibody followed by enhanced chemiluminescence detection.
3. Biotin: block the membrane with 0.5% (w/w) BSA and 0.1% (v/v) Tween-20 in PBS for 2 h. Incubate the blot at a 1:5000 dilution of streptavidin-HRP conjugate in 0.5% (w/w), 0.1% (v/v) Tween-20 in PBS for 1 h. Wash the blot five times with PBS containing 0.1% (v/v) Tween-20 for 10–15 min for each washing step. Develop the blot by chemoluminescense.

3.4. Synthesis of Probe A: AdaK(Bio)Ahx_3L_3VS (A) (see *Fig. 2*)

1. Fmoc-Leu-Wang resin (0.5 mmol) was elongated by standard Fmoc-based solid-phase peptide synthesis (SPPS) with Fmoc-Leu-OH, Fmoc-Ahx-OH, Fmoc-Ahx-OH, Fmoc-Ahx-Oh, Fmoc-Lys(biotin)-OH, and adamantaneacetic acid, each time 5 eq of the amino acid, 5 eq of PyBOP as the activating agent, and 6 eq of diisopropylethyl amine as the base.
2. After each condensation step, the N-terminal Fmoc protective group was removed by treatment with 20% piperidine in DMF.
3. Treatment of the immobilized peptide AdaK(bio)Ahx_3L_2OH with TFA/TIS/H_2O 95/2.5/2.5 v/v and removal *in vacuo* of the solvents, was followed by solution phase block-coupling with leucine vinyl sulfone TFA salt (1 eq) under the agency of HBTU (1 eq) and DiPEA (2.2 eq) in DMF.
4. After evaporation of the solvent, the residue was dissolved in ethyl acetate.
5. Precipitation of the product was accomplished by sonication for 2 min. The precipitate was filtered and washed with ethyl acetate, diethyl ether, and hexanes to afford the title compound A in 40% yield with a purity of approx 90%.
6. The peptide can be purified to homogeneity by reversed-phase HPLC or by silica gel chromatography using a gradient from 10% to 20% methanol in methylene chloride. MS: found 1287.9 $(M+H)^+$. For ^{1}H-NMR (DMSO-d6, 500 MHz) data, *see* **Note 5**.

3.5. Synthesis and Radiolabeling of Probe B: AdaY(^{125}I)Ahx_3L_3VS (B)

1. Ada$YAhx_3L_3$VS: synthesized as compound A, but with Fmoc-Tyr(OtBu)-OH instead of Fmoc-Lys(biotin)-OH. MS: found 1097.1 $(M+H)^+$. For ^{1}H-NMR (DMSO-d6, 500 MHz) data, *see* **Note 6**.

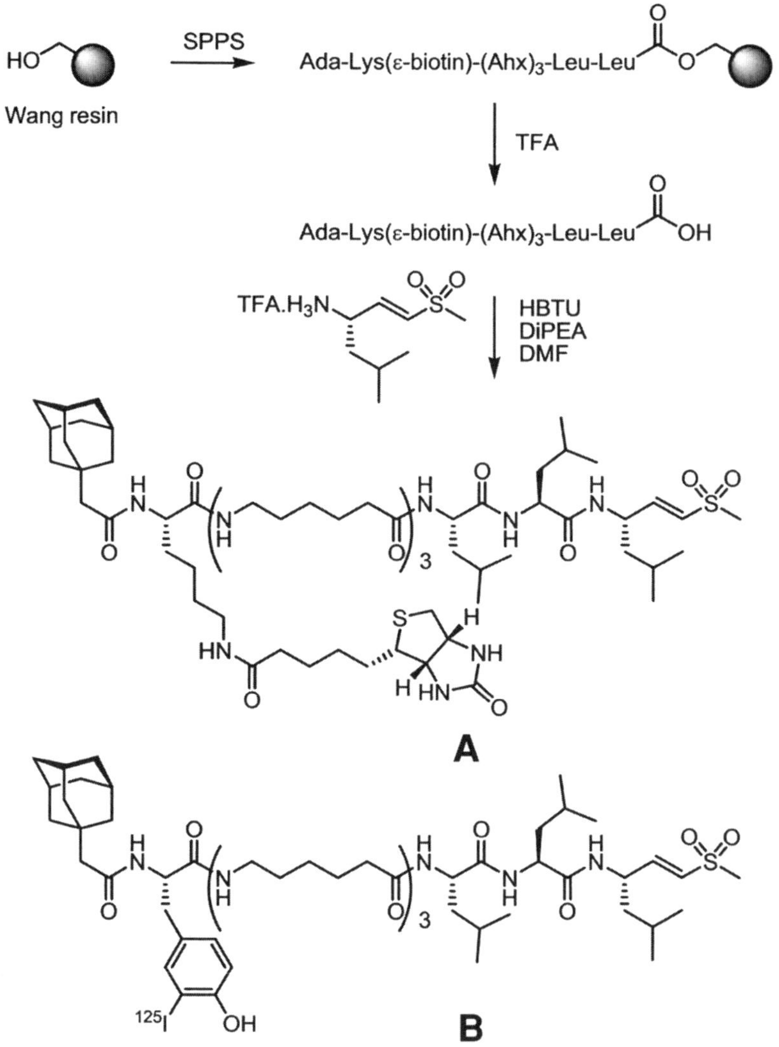

Fig. 2. Synthesis of AdaK(Bio)Ahx_3L_3VS (A) and chemical structure of AdaY(^{125}I)Ahx_3L_3VS (B).

2. Radiolabeling: Iodo-gen (100 μg; *see* **Note 3**) was dissolved in 50 μL phosphate buffer.
3. AdaYAhx_3L_3VS (20 μg) in acetonitrile (30 μL) was added and the mixture was agitated for 2 min. $Na^{125}I$ (10 μL aqueous solution, 1 mCi) was added to the mixture and agitated for 10 min.
4. The radioiodinated proteasome inhibitor was separated using a Sep-pak C_{18} column. Free iodine was washed away with phosphate buffer (pH 7.4), and radioiod-

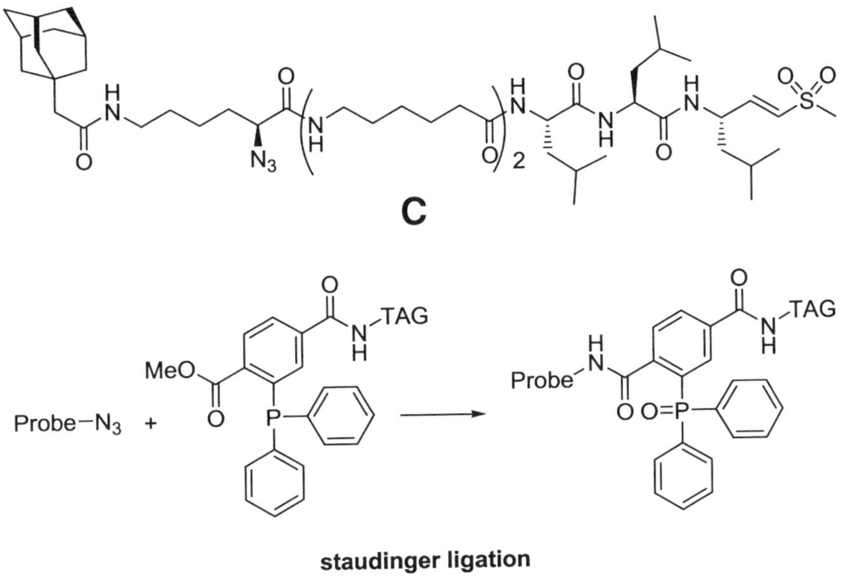

Fig. 3. Chemical structure of AdaAhx(α-N_3)Ahx_2L_3VS (C) and reaction scheme for the Staudinger ligation.

inated AdaY(^{125}I)Ahx_3L_3VS B was eluted using acetonitrile. One-milliliter fractions were collected and the fractions containing most radioactivity (typically 0.3–0.7 × 10^6 cpm/μL) were used for proteasome radiolabeling experiments.

3.6. Synthesis of Probe C: AdaAhx(α-N_3)Ahx_2L_3VS (C) (see Fig. 3)

1. Solid phase peptide synthesis was carried out essentially as described for Probe A.
2. Synthesis of Fmoc-Ahx(α-N_3)-OH: *N*-(α-Boc),*N*-(ε-Fmoc)-L-lysine (2.0 g, 4.2 mmol) was treated with 50% v/v TFA/CH_2Cl_2 for 40 min.
3. Solvents were removed *in vacuo*, yielding 1.9 g (3.9 mmol, 92%) of crude TFA-salt of *N*-(ε-Fmoc)-L-lysine.
4. The crude product was treated with trifluoromethanesulfonyl azide as described in **ref. *20***, followed by a modified work-up procedure.
5. After removal of the organic solvents, the aqueous slurry was acidified with 1 *M* HCl to pH 2.0 and extracted with EtOAc (4X). The combined organic phases were dried over $MgSO_4$, filtered, and concentrated *in vacuo*.
6. Purification of the crude product over silica gel (hexanes:EtOAc 2:1) yielded 1.4 g (3.5 mmol, 89%) of a white foam. Electrospray ionization (ESI)-MS: *m/z* (%) = 417.2 (100) [M+Na^+], 811.5 (20) [2M+Na^+]. For ^{1}H NMR (200 MHz, $CDCl_3$, 25°C, TMS) and ^{13}C NMR (50.1 MHz, $CDCl_3$, 25°C, TMS) data, *see* **Note 7**.

7. Next, Wang resin (1.0 g, 0.86 mmol) was condensed with Fmoc-Leu-OH (1.2 g, 3.4 mmol) in CH_2Cl_2 (25 mL) under the agency of diisopropylcarbodiimide (0.54 mL, 3.44 mmol) and a catalytic amount of 4-(dimethylamino)pyridine for 2 h.
8. The resin was filtered off, washed (3X CH_2Cl_2-MeOH, CH_2Cl_2, and Et_2O), and air dried. Loading was determined by quantification (ultraviolet) of Fmoc cleavage, and proved to be 0.66 mmol/g.
9. The resin, 260 mg (0.17 mmol), was elongated using standard Fmoc-SPPS to give resin-bound AdaAhx(α-N_3)Ahx_2L_2.
10. Treatment of the immobilized peptide with TFA/H_2O 95/5 v/v for 1 h, filtration, and removal of solvent *in vacuo* was followed by a solution phase block coupling with leucine vinyl sulfone TFA salt (1 eq) under the agency of HBTU (1 eq) and DiPEA (2.2 eq) in DMF. After evaporation of the solvent, the residue was dissolved in EtOAc. Precipitation of the product was accomplished by sonication. The precipitate was filtered and washed with EtOAc, Et_2O, and hexanes to yield 146 mg (0.15 mmol, 88%) of the title compound in 90% purity as judged by LC-MS.
11. Silica gel purification (0–10% MeOH/EtOAc) of 21 mg (22 µmol) of the crude product yielded 14 mg (14 µmol) of the title compound (58% overall yield) ready to use in biological experiments. LC-MS: $m/z = 974.9$ [$M+H^+$].

3.7. Synthesis of the Staudinger Ligation Handle

1. Fmoc Rink amide resin (128 mg, 100 µmol) was deprotected and *N*-(α-Boc),*N*-(ε-Mtt)-l-lysine (0.5 mmol, 312 mg) was coupled.
2. The Mtt protecting group was removed by treatment of the resin with 1% TFA in CH_2Cl_2 for 30 s. This treatment was repeated nine times, until no more yellow color was observed in the eluted solution. The resin was neutralized with 10% DiPEA in DMF.
3. Biotin (0.5 mmol, 122 mg) was coupled under the agency of PyBOP (1 eq) and DiPEA (1.2 eq) in DMF. After removal of the Fmoc group, the resulting free amine was condensed with 1 (*see* **Fig. 4**) ***(21)*** (0.5 mmol, 237 mg).
4. After washing of the immobilized peptide with dioxane (3X), the azide moiety was reduced by treatment with Me_3P (0.6 mL of a 1 *M* solution in toluene, 0.6 mmol, 6 eq) in dioxane/water (4/1 v/v, 2 mL) for 40 min ***(20)***, followed by dioxane washes (3X), and Fmoc-Ahx-OH (0.5 mmol, 180 mg) was coupled.
5. Of the obtained resin, 50 µmol was treated with piperidine as described previously to remove the Fmoc protecting group, and the resin was then washed with CH_2Cl_2 (3X).
6. Phosphine **2** (*see* **Fig. 4** and **ref.** ***22***) (90 mg, 0.25 mmol) was activated with EDC (48 mg, 0.25 mmol) and HOBt (41 mg, 0.3 mmol) in CH_2Cl_2 (2 mL) under an argon atmosphere for 5 min, and subsequently added to the resin.
7. The resin was agitated under argon atmosphere for 1 h, and the resin was washed (CH_2Cl_2, then DMF-MeOH alternating [3X], CH_2Cl_2-MeOH alternating [3X], and CH_2Cl_2), while keeping the resin under argon atmosphere.
8. The immobilized peptide was liberated from the resin by treatment with 50% v/v TFA/CH_2Cl_2 for 1 h. Evaporation of the solvents *in vacuo* followed by HPLC

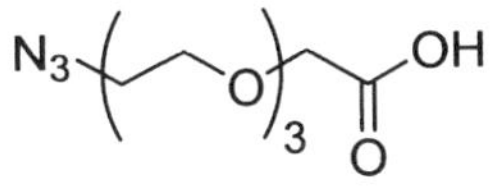

1

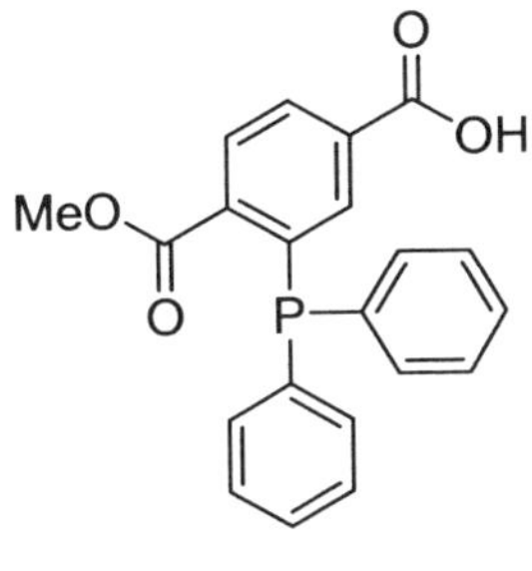

2

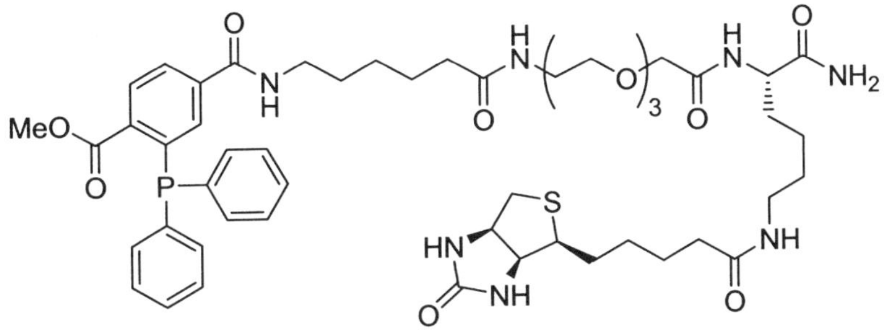

staudinger ligation handle

Fig. 4. Chemical structure of unusual building blocks 1 and 2 and the Staudinger ligation handle.

purification of the crude product (linear gradient in B: 25–55% B in three column volumes) yielded 8.0 mg (8 μmol, 16%) of a white solid. LC-MS: *m/z* calculated = 1020 [M+H$^+$], m/z found = 1020 [M+H$^+$].

9. The product was kept as aliquots of a stock solution (1.6 m*M*) in degassed DMSO at –80°C.

3.8. Synthesis of Probe D: DansylAhx$_3$L$_3$VS (D) (see Fig 5)

1. Dansyl-Ahx$_3$L$_2$-OH was synthesized on Wang resin using standard Fmoc-based solid-phase peptide synthesis protocols and purified by silica gel chromatography using a gradient of 10% → 20% of MeOH in CH_2Cl_2.
2. Dansyl-Ahx$_3$L$_2$-OH was coupled by *N*-(3-dimethylaminopropyl)-*N*9-ethylcarbodiimide hydrochloride (EDCI)-mediated condensation with (S,*E*)-5-methyl-1-(methylsulfonyl)hex-1-en-3-amine (leucine vinyl sulfone, LeuVS TFA salt) to afford D. MS(ESI): Calculated 990.5 $(M+H)^+$ 1012.5 $(M+Na)^+$. Observed 990.5 $(M+H)^+$ 1012.5 $(M+Na)^+$.
3. Hydrogenation and deprotection of Boc-LeuVS followed by condensation of the resulting (S)-5-methyl-1-(methylsulfonyl)hexan-3-amine Dansyl-Ahx$_3$L$_2$-OH

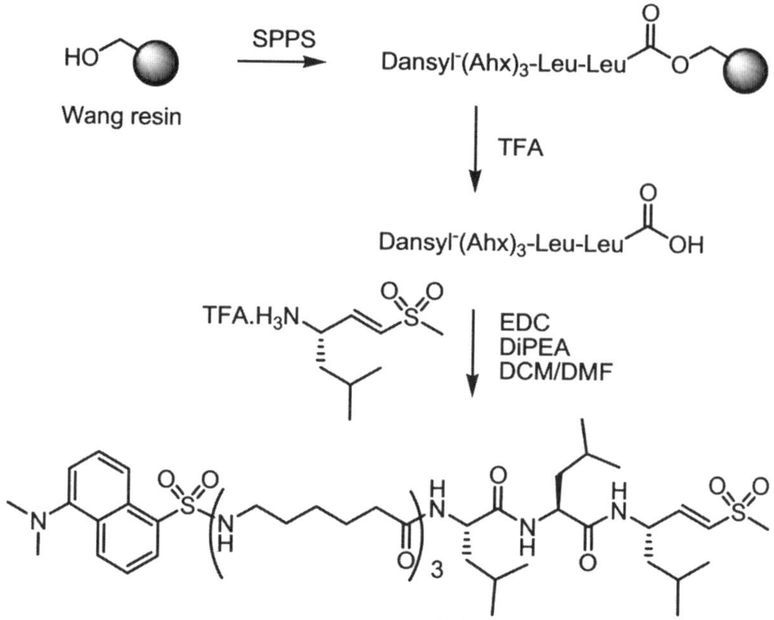

Fig. 5. Synthesis of DansylAhx$_3$L$_3$VS (D).

afforded the negative control (mock). ESI-MS: Calculated 992.5 $(M+H)^+$ 1014.5 $(M+Na)^+$. Observed 992.5 $(M+H)^+$ 1014.5 $(M+Na)^+$.

3.9. Proteasome Profiling Using AdaK(Bio)Ahx$_3$L$_3$VS (A)

1. Incubate cell lysates (25 µg of total protein) with 3 µg AdaK(Bio)Ahx$_3$L$_3$VS (A) for 2 h at 37°C.
2. Add 3X SDS sample buffer (1X final concentration).
3. Analyze by one-dimensional SDS-PAGE (12.5%).
4. Western blotting, probing with streptavidin-HRP, and chemoluminescence detection.
5. For 2D NEPHGE SDS-PAGE, add 220 µL of rehydration solution to 30 µL of samples and mix.
6. Subject to first-dimension isoelectric focusing using a IPGphor isoelectric focusing system (Amersham Pharmacia Biotech) and second-dimension SDS-PAGE (12.5%). (The results obtained were comparable to conventional 2D NEPHGE SDS-PAGE methods.)
7. Western blotting, probing with streptavidin-HRP, and chemoluminescence detection.

3.10. Proteasome Profiling Using AdaY(^{125}I)Ahx$_3$L$_3$VS (B)

1. Incubate cell lysates (25 µg of total protein) with AdaY(^{125}I)Ahx$_3$L$_3$VS (B) (0.5 × 10^6 cpm) for 2 h at 37°C.

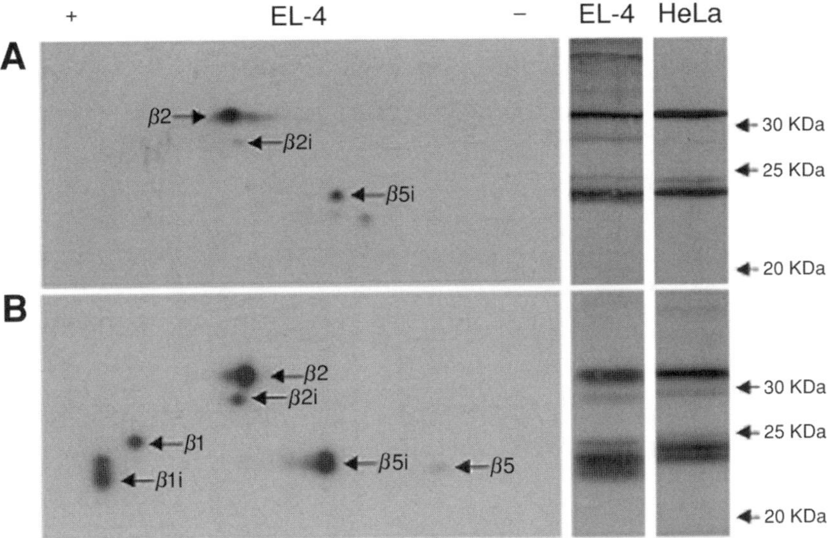

Fig. 6. 1D- and 2D gel analysis of EL-4 and HeLa cell lysates labeled with either AdaK(Bio)Ahx_3L_3VS (A) or AdaY(^{125}I)Ahx_3L_3VS (B).

2. Add 3X SDS sample buffer (1X final concentration), heat to 95°C for 3 min.
3. Analyze by one-dimensional SDS-PAGE (12.5%).
4. Autoradiography.
5. For 2D NEPHGE SDS-PAGE, add 220 µL of rehydration solution to 30 µL of samples and mix.
6. Subject to first-dimension isoelectric focusing using a IPGphor isoelectric focusing system (Amersham Pharmacia Biotech) and second-dimension SDS-PAGE (12.5%). (The results obtained were comparable with conventional 2D NEPHGE SDS-PAGE methods; *see* **Fig. 6**.)
7. Autoradiography.

3.11. Proteasome Profiling in Living Cells Using Two-Step Labeling With AdaAhx(α-N_3)Ahx_2L_3VS (C)

1. Incubate 5 × 10^6 EL-4 cells cultured in RPMI (Gibco, Invitrogen Corp.) supplemented with L-glutamine, FCS, penicillin, and streptomycin overnight with concentrations of AdaAhx(α-N_3)Ahx_2L_3VS C as indicated in **Fig. 7**.
2. Centrifuge and harvest the cells.
3. After glass-bead lysis, resuspend 10 µg of protein in 33 µL of lysis buffer and denatured by the addition of 2 µL of 20% SDS followed by brief boiling.
4. Incubate the denatured sample with 10 µL (100 µ*M* in DMSO/CH_3CN/H_2O 6/1/1 v/v/v) of Staudinger ligation handle for 1 h at 37°C.
5. Add 4X sample buffer.

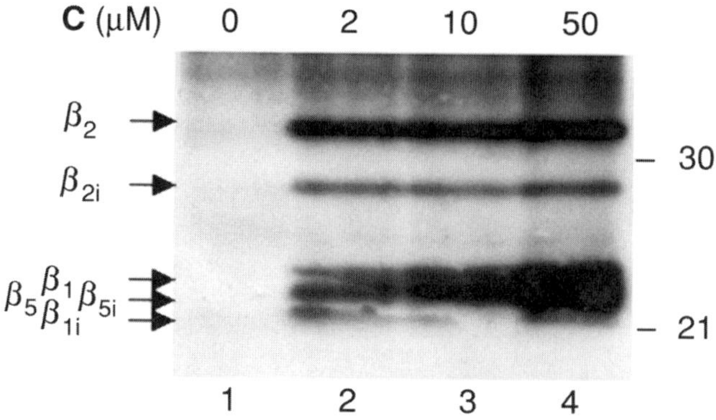

Fig. 7. 1D gel analysis of EL-4 cells labeled with AdaAhx(α-N_3)Ahx_2L_3VS (C).

6. Run SDS-PAGE (12.5%).
7. Western blotting, probing with streptavidin-HRP, and chemoluminescence detection.

3.12. Proteasome Profiling in Living Cells Using DansylAhx$_3$L$_3$VS (D) (see *Fig. 8*)

1. Incubate 10×10^6 EL4 cells in 10 mL RPMI 1640 with the indicated amounts of probe D or 10 μ*M* mock, at 37°C for 2 h.
2. Harvest cells by centrifugation.
3. Lyse cells by glass bead lysis or by incubating them for 30 min in NP-40 lysis buffer and centrifuge 5 min to remove membrane fractions, nuclei, and cell debris.
4. Determine protein content by the Bradford method.
5. Denature equal amounts (typically 60 μg) of protein by boiling in reducing sample buffer.
6. Run SDS-PAGE (12.5%).
7. Western blotting, probing with rabbit anti-dansyl primary antibody, goat anti rabbit HRP-labeled secondary antibody, and chemoluminescence detection.

4. Notes

1. Iodoacetonitrile was passed over a basic alumina plug prior to use.
2. A higher Tween-20 concentration (0.1%) should be used when probing with streptavidin-HRP because of high nonspecific background adsorption.
3. A film of iodo-gen was deposited on the reaction vial by dissolving in choloroform followed by solvent evaporation.
4. We found that 0.5% BSA in TBS-T as blocking and incubation solution, followed by extensive washing with TBS-T (5 × 5 min with 5 to 10 mL), gives optimal results. We experienced some variation between different antibody lot numbers.

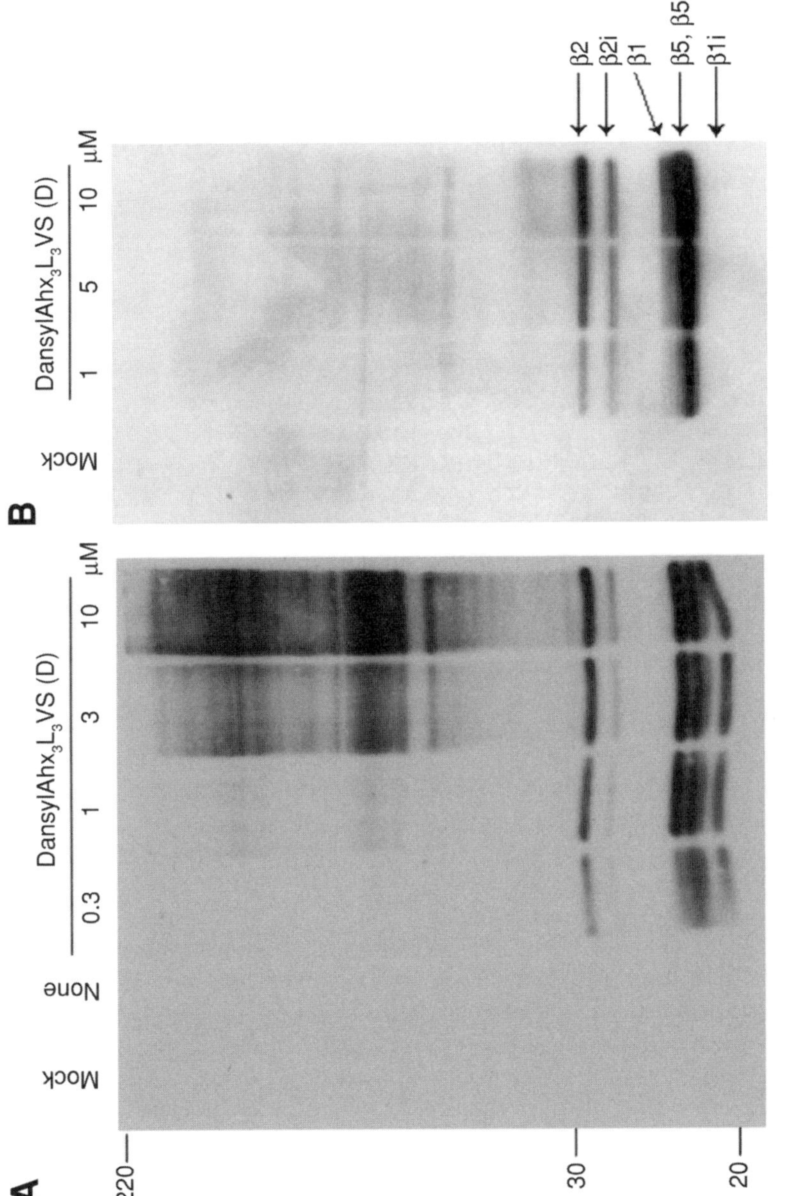

Fig. 8. 1D gel analysis of A: EL-4 cell lysates and B: EL-4 cells labeled with DansylAhx$_3$L$_3$VS (D).

5. Probe A: AdaK(Bio)Ahx$_3$L$_3$VS (A): MS: found 1287.9 $(M+H)^+$; ^{1}H-NMR (DMSO-d6, 500 MHz): δ 8.11 (d, 1H, J 8.5 Hz), 7.99–7.93 (m, 2H), 7.84 (d, 1H, J 8.0 Hz), 7.74–7.68 (m, 2H), 7.07 (d, 1H, J 10.5 Hz), 6.68–6.56 (m, 3H), 4.57–4.50 (m, 1H), 4.31–4.10 (m, 4H), 4.00–3.94 (m, 1H), 3.08–2.94 (m, 16H), 2.14–2.08 (m, 2H), 2.04–1.98 (m, 4H), 1.88 (s, 2H), 1.62–1.16 (m, 54H), 0.91–0.80 (m, 18H).
6. Probe B: AdaYAhx$_3$L$_3$VS (B): MS: found 1097.1 $(M+H)^+$. For ^{1}H-NMR (DMSO-d6, 500 MHz): δ 8.12 (d, 1H, J 9.0 Hz), 7.98–7.92 (m, 3H), 7.85–7.80 (m, 2H), 7.75–7.68 (m, 2H), 6.99–7.97 (m, 2H), 6.67–6.58 (m, 4H), 4.58–4.50 (m, 1H), 4.41–4.35 (m, 1H), 4.28–4.20 (m, 2H), 3.05–2.95 (m, 9H, 2.81 (dd, 1H, J 5.0 and 13,5 Hz), 2.62 (dd, 1H, J 9.8 and 13.5 Hz), 2.14–2.04 (m, 2H), 2.02–1.96 (m, 4H), 1/80 (s, 2H), 1.60–1.24 (m, 42H), 0.91–0.80 (m, 18H).
7. Fmoc-Ahx(a-N$_3$)-OH: ESI-MS: *m/z* (%) = 417.2 (100) $[M+Na^+]$, 811.5 (20) $[2M+Na^+]$; ^{1}H NMR (200 MHz, $CDCl_3$, 25°C, TMS): δ = 7.75 (d, ^3J (H,H)=7.3 Hz, 2H; ArH), 7.57 (d, ^3J (H,H)=7.3 Hz, 2H; ArH), 7.40–7.25 (m, 4H; ArH), 4.41 (d, ^3J (H,H)=6.6 Hz, 2H; $CHCH_2O$), 4.19 (t, ^3J (H,H)=6.6 Hz, 1H; $CHCH_2O$), 3.92 (t, ^3J (H,H)=6.7 Hz, 1H; αCH), 3.18 (m, 2H; εCH_2), 1.7–1.3 ppm (m, 6H; βCH_2, γCH_2, δCH_2); ^{13}C NMR (50.1 MHz, $CDCl_3$, 25°C, TMS): δ = 174.1, 156.8, 143.5, 140.9, 127.1, 126.7, 124.6, 119.6, 66.4, 62.1, 60.4, 46.7, 30.6, 28.8, 22.4 ppm.

References

1. Williams, A., Peh, C. A., and Elliott, T. (2000) The cell biology of MHC class I antigen presentation. *Tissue Antigens* **59,** 3–17.
2. Robinson, J. H. and Delvig, A. A. (2002) Diversity in MHC class II antigen presentation. *Immunology* **105,** 252–262.
3. Yewdell, J. W., Reits, E., and Neefjes, J. (2003) Making sense of mass destruction: quantitating MHC class I antigen presentation. *Nat. Rev. Immunol.* **3,** 952–961.
4. Lehner, P. J. and Cresswell, P. (2004) Recent developments in MHC-class-I-mediated antigen presentation. *Curr. Opin. Immunol.* **16,** 82–89.
5. Wilson, N. S. and Villadangos, J. A. (2005) Regulation of antigen presentation and cross-presentation in the dendritic cell network: facts, hypothesis, and immunological implications. *Adv. Immunol.* **86,** 241–305.
6. Rock, K. L. and Goldberg, A. L. (1999) Degradation of cell proteins and the generation of MHC class I-presented peptides. *Annu. Rev. Immunol.* **17,** 739–779.
7. Dahlmann, B., Ruppert, T., Kloetzel, P. M., and Kuehn, L. (2001) Subtypes of 20S proteasomes from skeletal muscle. *Biochimie* **83,** 295–299.
8. Driscoll, J., Brown, M. G., Finley, D., and Monaco, J. J. (1993) MHC-linked LMP gene products specifically alter peptidase activities of the proteasome. *Nature* **365,** 262–264.
9. Gaczynska, M., Rock, K. L., and Goldberg, A. L. (1993) Gamma-interferon and expression of MHC genes regulate peptide hydrolysis by proteasomes. *Nature* **365,** 264–267.
10. Adam, G. C., Cravatt, B. F., and Sorensen, E. J. (2001) Profiling the specific reactivity of the proteome with non-directed activity-based probes. *Chem. Biol.* **8,** 81–95.

11. Adam, G. C., Sorensen, E. J., and Cravatt, B. F. (2002) Proteomic profiling of mechanistically distinct enzyme classes using a common chemotype. *Nat. Biotechnol.* **20,** 805–809.
12. Ovaa, H., van Swieten, P. F., Kessler, B. M., Leeuwenburgh, M. A., Fiebiger, E., van den Nieuwendijk, A. M., et al. (2003) Chemistry in living cells: detection of active proteasomes by a two-step labeling strategy. *Angew. Chem. Int. Ed. Engl.* **42,** 3626–3629.
13. Bogyo, M., McMaster, J. S., Gaczynska, M., Tortorella, D., Goldberg, A. L., and Ploegh, H. (1997) Covalent modification of the active site threonine of proteasomal beta subunits and the *Escherichia coli* homolog HslV by a new class of inhibitors. *Proc. Natl. Acad. Sci. USA* **94,** 6629–6634.
14. Bogyo, M., Shin, S., McMaster, J. S., and Ploegh, H. L. (1998) Substrate binding and sequence preference of the proteasome revealed by active-site-directed affinity probes *Chem. Biol.* **5,** 307–320.
15. Greenbaum, D., Medzihradszky, K. F., Burlingame, A., and Bogyo, M. (2000) Epoxide electrophiles as activity-dependent cysteine protease profiling and discovery tools. *Chem. Biol.* **7,** 569–581.
16. Liu, Y., Patricelli, M. P., and Cravatt, B. F. (1999) Activity-based protein profiling: the serine hydrolases *Proc. Natl. Acad. Sci. USA* **96,** 14,694–14,699.
17. Kessler, B. M., Tortorella, D., Altun, M., Kisselev, A. F., Fiebiger, E., Hekking, B. G., et al. (2001) Extended peptide-based inhibitors efficiently target the proteasome and reveal overlapping specificities of the catalytic beta-subunits. *Chem. Biol.* **8,** 913–929.
18. Berkers, C. R., Verdoes, M., Lichtman, E., Fiebiger, E., Kessler, B. M., Anderson, K. C., et al. (2005) Activity probe for in vivo profiling of the specificity of proteasome inhibitor bortezomib. *Nat. Methods* **2,** 357–362.
19. Palmer, J. T., Rasnick, D., Klaus, J. L., and Bromme, D. (1995) Vinyl sulfones as mechanism-based cysteine protease inhibitors. *J. Med. Chem.* **38,** 3193–3196.
20. Lundquist, J. T. and Pelletier, J. C. (2001) Improved solid-phase peptide synthesis method utilizing alpha-azide-protected amino acids. *Org. Lett.* **3,** 781–783.
21. Jeong, S. W. and O'Brien, D. F. (2001) Synthesis of a polymerizable metal-ion-chelating lipid for fluid bilayers. *J. Org. Chem.* **66,** 4799–4802.
22. Kiick, K. L., Saxon, E., Tirrell, D. A., and Bertozzi, C. R. (2002) Incorporation of azides into recombinant proteins for chemoselective modification by the Staudinger ligation. *Proc. Natl. Acad. Sci. USA* **99,** 19–24.

5

Two-Dimensional Difference Gel Electrophoresis

Terence L. Wu

Summary

The two-dimensional (2D) polyacrylamide gel-based approach to protein profiling has been successful because it is an accessible, inexpensive, and powerful tool for the analysis of global patterns of protein expression. All protein spots that are resolved and detected within the 10^4 to 10^5 dynamic range of gel capacity can be studied qualitatively and quantitatively in relation to each other, and viewed as a single image. Two-dimensional difference gel electrophoresis (DIGE) has strengthened the 2D platform by allowing the detection and quantitation of differences between samples resolved on the same gel, or across multiple gels, when linked by an internal standard. The technology is based on modification of protein extracts with fluorescent cyanine dyes, which have distinct excitation and emission spectra and are migration (charge and size) matched so that the same protein labeled with any of the dyes (Cy2, Cy3, Cy5) will migrate to the same position within a 2D gel, greatly enhancing reproducibility.

It is becoming clear that each technology that is currently available for quantifying differential protein expression has its own weaknesses and strengths, and that multiplatform, integrated approaches will be necessary to provide the most complete analysis of any given proteome. We believe that DIGE is, and will remain in the future, a key front-end proteomic tool that will strongly complement other protein-profiling technologies.

Key Words: Difference gel electrophoresis (DIGE); fluorescent labeling; differential proteomic analysis.

1. Introduction

Two-dimensional gel electrophoresis is a powerful and practical proteomics tool, capable of simultaneously separating and resolving complex protein mixtures and enabling the visualization and identification of several thousand proteins on a single gel. This resolution is achieved by separating proteins based on their isoelectric points in the first dimension and then using sodium dodecyl sulfate (SDS)-polyacrylamide gel electrophoresis (PAGE) to separate them based on their apparent size in the second dimension. The isoelectric focusing

From: *Methods in Molecular Biology, Vol. 328: New and Emerging Proteomic Techniques*
Edited by: D. Nedelkov and R. W. Nelson © Humana Press Inc., Totowa, NJ

(IEF) of the first dimension is performed using immobilized pH gradient (IPG) strips and ampholines *(1)*, with a range of separation extending from a broad range (pH 3.0-11.0 is currently available) on a single gel to more refined, single pH unit separations. The second-dimension SDS-PAGE gels can be poured with either single or gradient concentrations of acrylamide, with a range of resolution extending from below 10 kD to more than 200 kD. Protein spots of interest can be excised from gels and then digested *in situ* with trypsin or other proteases. These digests can be extracted from the gels and then subjected to a range of different mass-spectrometric technologies. The resulting mass data are then used to search protein, genomic, expressed sequence tag (EST), and other species-specific databases to identify proteins present in each selected spot. In addition to identifying proteins present in each spot of interest, protein isoforms and posttranslational modifications such as phosphorylation and glycosylation can also frequently be detected and studied by 2D gel electrophoresis. By comparing spot patterns and intensities from different samples, changes in the level of expression of individual proteins can be detected and quantified, thus permitting the identification of biomarkers associated with specific pathologic or physiologic states of a cell or tissue *(2)*. One limitation of traditional comparative 2D gel electrophoresis is the high level of gel-to-gel variation in spot patterns, which often confounds the identification of the same protein-containing spots in different gels and which also often results in different boundaries within which the "same" spot may be integrated on two different gels. The difficulty in reproducing spot patterns on different gels makes it difficult to distinguish true biological from experimental variation.

2D difference gel electrophoresis (2D-DIGE) *(3,4)* utilizes up to three mass- and charge-matched, spectrally resolvable fluorescent dyes (Cy2, Cy3, and Cy5) to label a control and two different protein samples in vitro prior to 2D gel electrophoreis. Currently, there are two different protein-labeling chemistries for DIGE. The first approach uses relatively hydrophobic cyanine dyes with NHS-ester reactive groups. To prevent dye-labeled proteins from becoming too insoluble, these "minimal labeling" dyes are designed to label just 1–2% of available lysines. The quantitative and qualitative reproducibility challenges inherent in trying to align and similarly integrate the same 1000 or more spots in two different 2D electrophoreis gels are overcome by running the control, experimental, and pooled internal standard (if used in a multiplexed experiment) samples together through the same first- and second-dimension gels. While these samples are imaged separately, each of the resulting images can be perfectly overlaid, because the same spots from all three samples migrate together. As a result, the confidence and the error limits within which the protein expression level changes can be quantified between the samples are substantially improved. With sample multiplexing, one dye (Cy2) is used to label a pooled internal standard comprised of equal portions of each sample in an experiment. This internal standard is run on each gel of the experiment and is used to normalize protein abundance measurements across multiple gels in the analysis. In the case of large-volume spots, this approach allows the detection of relative changes in protein expression that are as little as 1.2-fold ***(2)***. With current 2D image analysis software suites such as DeCyder (GE/Amersham), robotics may be implemented, which substantially automates spot characterization and integration, excision of spots that meet user-defined criteria for differential expression, and subsequent in-gel digestion. DIGE fluorescence-based detection has a large dynamic range of 10^4-10^5, and the dye sensitivity is capable of detecting 0.25 ng to 1 ng of sample, thus enabling the detection of relatively low copy-number proteins. As a result of this excellent sensitivity, DIGE can be used to analyze relatively small amounts of even very complex cell extracts. For example, when minimal NHS-ester CyDyes are used, as little as 50 μg of total cell protein extract per sample can be subjected to DIGE. About 20% of the proteins in this sample would be expected to be labeled via a covalent amide linkage on a single lysine (i.e., approx 1–4% of all lysine residues are labeled in all proteins in the sample).

Even less sample is needed for DIGE analysis when using the more recently introduced "saturation" CyDyes, which enable the use of as little as 5 μg protein per sample. This approach is useful when probing for differential protein expression with scarce amounts of sample; however, many of the detected spots will be too low in quantity to identify by spot picking and liquid chromatography (LC)-tandem mass spectrometry (MS/MS) analysis. As mentioned previously, the target for these saturation dyes are all cysteine-containing proteins. Because more than 96% of the proteins that have been identified in the human proteome contain cysteine ***(5)***, the resulting coverage is very comprehensive. The saturation CyDyes have a maleimide reactive group designed to label all accessible cysteines on all proteins with a covalent thioether linkage, resulting in higher sensitivity of detection vs minimal labeling CyDyes. One limitation is that currently only two saturation dyes are available (Cy3 and Cy5), thus complicating gel multiplexing, as more gels must be utilized in an experiment. In order to use an internal standard with the saturation dyes, one dye (Cy3) is used for the pooled internal standard (consisting of equal portions of all the samples in the experiment), which is run on all gels in an experimental series. The other (Cy5) dye is used for the test samples. The protein abundance is then expressed as a normalized ratio relative to spots from the internal standard.

The sensitivity of the CyDye minimal labeled proteins is comparable to silver or Sypro Ruby stains ***(4,6)***, and the saturation dyes are even more sensitive ***(7)***, at least for those proteins containing multiple cysteine residues. Using current mass spectrometry (MS) technology, it should in theory be possible to identify even the fainter spots (approx 1 ng/spot or approx 20 fmols of a 50-kD protein), as the sensitivity of some MS platforms now lies in the attomole range. However, the practical limitations of not being able to quantitatively pick the entire spot (i.e., with automated spot pickers having a fixed diameter picker) and inefficiencies of the digestion/extraction procedures lead to overall recoveries that are well below 100%, which often confound the MS-based identification of less-intense DIGE spots. Loading additional unlabeled sample increases the probability of extracting sufficient amounts of protein to identify by MS, but risks loss of resolution by overloading. It is of note that the CyDye labeling process is a mass action event and that the dynamic range of approx 10^4 to 10^5 afforded by the dyes would necessarily restrict labeling to only the corresponding concentration range of the most abundant proteins in the mixture. For example, in human sera, the dynamic range of protein abundance extends to at least over a 10^{10} range in protein concentrations, with the approx 40 mg/mL serum albumin being the most abundant protein ***(8,9)***. In this instance the 10^4 to 10^5 range of DIGE would theoretically allow this technology to detect proteins expressed over the 0.4 or 4.0 μg/mL to 40 mg/mL range. To detect less-abundant proteins, it would be necessary to either prefractionate (sera) or deplete the more abundant proteins and concentrate the samples prior to labeling.

The power of DIGE technology over traditional 2D methods has been demonstrated for a wide variety of biological applications (e.g., ***10–16***), including studies related to the mitochondrial proteome from mouse heart ***(17)***, inner medullary collecting duct proteins from rat kidney cells ***(18)***, instantly released wound and immune proteins from *Drosophila* hemolymph ***(19)***, and studies related to breast ***(20)***, esophageal ***(21)***, gastric ***(22)***, and colon cancer ***(23)***. DIGE technology is a strong complement to other proteomic methods such as isotope-coded affinity tagging (ICAT), mass tagging, 2D-LC, peptide protein biomarker discovery, array technologies, and functional genomic approaches such as metabolomic and transcriptomic studies.

2. Materials

CyDyes for both minimal labeling and saturation labeling are from GE Healthcare (formerly Amersham Biosciences, Piscataway, NJ). For both minimal and saturation labeling lysis buffers, other detergents may be used, in place of or including CHAPS, to increase the solubility of membrane or very hydrophobic target proteins. It may be difficult or impractical to solubilize all samples in the standard lysis buffer; therefore, compatibility of other reagents with CyDye labeling is an issue that must be empirically tested on a per case basis. References to H_2O/water refer to double-distilled 18.2 MΩ H_2O, with organic content less than 5 ppb. The detection strategies employed here, from

Table 1
Effect of Other Reagents on Labeling by Minimal CyDyes *(24)*

Detergents	Concentrations	Effect on labeling
ASB-14	Up to 2%	Slight reduction in labeling
Triton X-100	If used at 1%	17% reduction in labeling
NP-40	Up to 1%	No effect on labeling
SDS	Up to 1%	No effect on labeling
Reducing agents		
DTT	2 mg/mL	Slight reduction in labeling
	5 mg/mL	2X reduction in labeling
TCEP	1 m*M* to 0.5 m*M*	Slight reduction in labeling

Avoid the use of amines and ampholines in the minimal dye lysis/labeling buffer, because these can cause a substantial reduction in labeling efficiency. The poor nucleophilicity of the Tris amine group relative to that of the epsilon primary amine of lysine enables the use of Tris as a buffer without compromising the efficiency of CyDye DIGE Fluor minimal dye labeling. SDS, sodium dodecyl sulfate; DTT, dithiothreitol; TCEP, Tris(2-carboxyethyl)phosphine hydrochloride.

the fluorescent dyes to the MS instrumentation used in the final analysis, are extremely sensitive, so appropriate precautions must be taken to prevent contamination of the samples and gels by dust, keratin, and the like.

2.1. Sample Preparation

2.1.1. Solubilization With Lysis Buffer for Minimal Dyes

1. Lysis/labeling buffer: 7 *M* urea, 2 *M* thiourea, 4% (w/v) CHAPS (Calbiochem), 30 m*M* Tris (pH 8.5) at 4°C. Check that the final solution pH is 8.5 at 4°C, to ensure good labeling efficiency. Dispense into approx 1-mL aliquots and store at -80°C, stable for several weeks.
2. Quenching solution for minimal CyDyes: 10 m*M* l-lysine in water.
3. Rehydration buffer: 7 *M* urea, 2 *M* thiourea, 4% (w/v) CHAPS, 2 mg/mL dithiothreitol (DTT), 1% Pharmalyte (various pH ranges are available), in water. Approximately 1-mL aliquots may be stored for several weeks at –20°C and longer at –70°C (*see* **Table 1** for comments on the effect of other reagents on labeling by minimal CyDyes).

2.1.2. Solubilization With Lysis Buffer for Saturation Dyes

1. Lysis/labeling buffer: 7 *M* urea, 2 *M* thiourea, 4% (w/v) CHAPS (Calbiochem), 30 m*M* Tris (pH 8.0) at 37°C. For the saturation dyes, make the final volume up to 25 mL with H_2O, and ensure that the stock pH of the Tris is 8.0 at 37°C, (or 8.3 at

25°C), to ensure good labeling efficiency. Dispense into approx 1-mL aliquots and store at -80°C, at which temperature the buffer is stable for several weeks.
2. 2X sample buffer (reaction stopping solution for saturation dyes): 7 *M* urea, 2 *M* thiourea, 4% (w/v) CHAPS, 2% (v/v) Pharmalyte (GE/Amersham, pH 3.0–10.0, or other appropriate range, depending on the IPG strip used), 1 mg/mL DTT, water.
3. Rehydration buffer: 7 *M* urea, 2 *M* thiourea, 4% (w/v) CHAPS, 1 mg/mL DTT, 1% Pharmalyte (various pH ranges available), water. Aliquots may be stored for several weeks at –20°C and longer at –70°C (*see* **Table 2** for comments on the effect of other reagents on labeling by saturation CyDyes.)

2.2. (Optional) Cleanup of Sample by Protein Precipitation Using the GE 2D Clean-up Kit (cat. no. 80-6486-60, GE/Amersham) (see Note 1)

1. Precipitant: renders proteins insoluble.
2. Co-precipitant: contains reagents that co-precipitate with proteins and enhances their removal from solution.
3. Wash buffer: used to remove nonprotein contaminants from the protein precipitate.
4. Wash additive: solution contains a reagent that promotes rapid and complete resuspension of the sample proteins.

2.3. Isoelectric Focusing (First Dimension)

1. The IEF apparatus is the IPGphor Isoelectric Focusing System from GE/Amersham, with an integrated temperature-regulating system and programmable power supply.
2. The 24-cm ceramic strip holders (80-6470-07, GE/Amersham) are used most often.
3. Equilibration tubes (80-6467-79, GE/Amersham).
4. Mineral oil (Dry Strip Cover Fluid, 17-1335-01, GE/Amersham).
5. IPG strips (Immobiline Dry Strip Gels, available in a wide range of pH ranges from GE/Amersham).
6. Immobiline Drystrip Reswelling Tray (80-6465-32, GE/Amersham).
7. Electrode strips (GE/Amersham).
8. IPG Buffer (ampholyte), of the appropriate range to match the IPG strip, both obtained from GE/Amersham.
9. Equilibration buffer (1): reducing buffer: 6 *M* urea, 10 m*M* Tris-HCl (pH 6.8) at room temperature, 30% (v/v) glycerol, 1% (w/v) SDS. Add DTT to a final concentration of 0.5% (w/v); use 10 mL/IPG strip.
10. Equilibration buffer (2): alkylating buffer-substitute iodoacetamide at a final concentration of 45 mg/mL in place of DTT; use 10 mL/IPG strip.

2.4. SDS-PAGE (Second Dimension)

1. The gel apparatus used is the Ettan DALTtwelve system, consisting of a 12-slot vertical slab separation unit, integrated 200-W, 600-V, 1.0-A programmable power unit, and a self-contained temperature-control and circulation unit (GE/Amersham).
2. Gel caster (no. 80-6467-22, with 14-gel capacity-GE/Amersham).
3. Low-fluorescence glass plates (80-6442-14, GE/Amersham).
4. Bind silane (17-1380-01, GE/Amersham).

Table 2
Effect of Other Reagents on Labeling by Saturation CyDyes *(24)*

Detergents	Concentrations	Effect on labeling
ASB-14	Up to 1%	No effect except that labeling conditions for saturation dyes (i.e., TCEP concentration) may have to be optimized when using different detergent combinations.
Triton X-100	Up to 4%	
NP-40	Up to 4%	
SDS	Up to 4%	
Reducing Agents		
DTT	1 mg/mL = 6.4 m*M*	Thiol-containing reagents will react with the saturation dyes; therefore, the amount of TCEP and dye may need to be increased accordingly.
β-mercaptoethanol	10 m*M*	
Other Reagents		
Amines	30 m*M*	10% reduction in labeling
Ampholines	1%	10% reduction in labeling

Effect of other reagents on both minimal and saturation labeling	
Salts	No effect if kept lower than 10 m*M* for rehydration loading, and less than 50 m*M* for cup loading
Protease inhibitors	Compatibility varies; see additional information in the Amersham/GE DIGE CyDye user guide.
Nuclease: (Benzonase, EMD Chemicals)	May be used at manufacturers suggested concentration, but these reagents may produce an artifact spot on the gel

SDS, sodium dodecyl sulfate; DTT, dithiothreitol; TCEP, Tris(2-carboxyethyl)phosphine hydrochloride; DIGE, difference gel electrophoresis.

5. Repel silane (17-1332-01, GE/Amersham).
6. Reference markers (18-1143-34, GE/Amersham).
7. Reagents include premixed acrylamide-bis solution (40% acrylamide , 3% bis-acrylamide), low-melting-point agarose, ammonium persulfate (APS), bromophenol blue (BPB), *n*-butanol glycine, Tris-HCl base, SDS, and *N,N,N′,N′*-tetramethylethylene-diamine (TEMED) of appropriate grade (electrophoresis or better). These reagents are available from Sigma-Aldrich (St. Louis, MO), GE/Amersham, and from other companies. Consistent use of reagent suppliers and lot numbers within a given set of experiments is recommended to reduce variability due to formulation differences. Note that precast gels, utilizing low-fluorescence glass plates in a customizable variety of concentrations, and gradients of high consistency, have been recently introduced by Jule, Inc. (Milford, CT).
8. Running buffer (pH 8.8): 30 g (25m*M* final) Tris-HCl base, 144 g (192 m*M* final) glycine, 10 g (0.01% final) SDS, H_2O to 10 L final volume.

9. Low melting point agarose solution: 0.5% (w/v) dissolved in 1X running gel buffer with a final concentration of 0.02% BPB.
10. Gel solution (for six 25 cm × 20 cm gels): 148.8 mL (40% acrylamide 3% bis-acrylamide); 120.0 mL 1.5 *M* Tris (pH 8.8) at 20°C; 4.8 mL 10% SDS; 204 mL H_2O.
11. Ammonium persulfate: 10% solution in water, make fresh prior to use.
12. Displacement solution: 50% (v/v) glycerol, 25% (v/v) 1.5*M* Tris (pH 8.8), water, trace amount of BPB. Approximately 100 mL/cast is needed.
13. Water-saturated iso-butanol. Shake well and let separate into phases before use.

2.5. Gel Scanning and Analysis

1. Images that are scanned using a Typhoon 9410 Variable Mode Imager (GE/Amersham) are output as GEL files, a variation of the 16-bit TIFF format, and are purely grayscale image files with one bit per pixel. The images have a dynamic range of 1-100,000 possible levels of signal resolution. A square-root compression algorithm is used to fit the potential 10^5 levels of image into the $2^{16} = 65,536$ actual available levels. The algorithm also provides higher signal resolution at the low end, where small changes in signal are critical. The emission (em)/excitation (ex) wavelength (nanometers) settings respectively are: for Cy2 = 520 em (40 bandpass)/488 ex, Cy3 = 580em, (30 bandpass)/532 ex, Cy5 = 670 em (30 bandpass)/633 ex, Sypro Ruby = 610 em (30 bandpass)/532 ex, Deep Purple = 610 em (30 bandpass)/457 ex.
2. DeCyder Differential Analysis Software (GE/Amersham) is an automated image analysis program suite that enables the detection, quantitation, matching, and analysis of 2D gels run with CyDye-labeled protein samples.
3. Poststain analysis is conducted using a high-sensitivity fluorescent dye such as Sypro Ruby (Molecular Bioprobes) or Deep Purple (Amersham/GE), which approximately matches the sensitivity of the CyDyes.
4. Fixer: 50% methanol, 10% acetic acid, H_2O.
5. Destain: 10% methanol, 7% acetic acid, H_2O.

2.6. Spot Picking, Tryptic Digestion

1. An Ettan Dalt spot-picking robot (GE/Amersham) interprets spot-picking coordinates generated through the DeCyder analysis program and transfers picked spots (gel plugs of 1.4-2.0 mm fixed diameter × 1 mm thick) into 96-well polypropylene plates.
2. An Ettan Dalt in-gel protease-digesting robot (GE/Amersham) is employed to wash, rinse, dehydrate, rehydrate, and digest the gel plugs with trypsin (Promega, cat. no. V5111, 20 μg/vial).

3. Methods

3.1. Sample Preparation and Labeling

3.2.2. Labeling With Minimal Dyes

With technical and manufacturing advances enabling excellent run-to-run reproducibility, it has become even more obvious that sample preparation remains the single most important and limiting aspect of the 2D-DIGE experimental approach. The potential sources and numbers of protein lysate samples that may be analyzed using 2D-DIGE technology is limited by the ability to disaggregate, fully solubilize, and maintain in solution the desired proteins in the sample pool in an appropriate sample buffer. The sample buffer must allow efficient, unbiased protein labeling while still maintaining conditions that allow entry and running on the first- and second-dimension gels (*see* **Notes 1–3** for comments about interfering chemical compounds and other, biological contaminants). As mentioned in the introduction, the large (10^4 to nearly 10^5) dynamic range of the DIGE platform allows detection of proteins that are present at concentrations that correspond to as little as 0.01 to 0.001% of the most abundant sample proteins. Nonetheless, the estimated 10^{10} dynamic range of human sera, and probably that of many other human tissue and fluid samples, precludes DIGE detection of the large number of lower-abundance proteins that are present at concentrations less than 0.001% of that of the most abundant protein. Detection of these lower-abundance proteins will require prefractionation of the sample and/or selective depletion of high-abundance proteins (*see* **Note 4**).

The recommended lysis/labeling buffer (*see* **Note 5**) for the DIGE minimal dye labeling system contains 7 *M* urea, 2 *M* thiourea, 4% CHAPS in 30 m*M* Tris (pH 8.5) at 4°C, and is used as a starting point for all sample types using these dyes. For the saturation CyDyes, the starting lysis/labeling buffer is 7 *M* urea, 2 *M* thiourea, 4% CHAPS in 30 m*M* Tris (pH 8.0) at 37°C. As mentioned, other reagent combinations may be necessary to accommodate the source material; however, these will have to be individually optimized for each sample to ensure compatibility with both the labeling and gel running procedures. In general, all procedures beginning at the moment of sample lysis should be kept as cold as possible without allowing reagent or sample precipitation (usually on ice or at 4°C) (*see* **Note 6**). Ideally, the samples should be labeled immediately and stored as cold as possible, preferably at –80°C. Freeze thawing in the lysis/labeling buffer may result in differential protein precipitation for some samples, so caution is advised. The same caution is advised for prolonged storage, even at –80°C. Although no single method can be applied universally, the following outline is for a frequently encountered sample of cultured cells grown in flasks with media containing serum.

1. Grow cells to the desired density. Wash attached cells several times with media without serum or phosphate-buffered saline that has been prewarmed to the incubation temperature. Scrape, or if necessary, trypsinize cells into a 15-mL conical tube (*see* **Note 7**).
2. Resuspend cell pellets in media without serum, again at the incubation temperature. Pellet the cells and repeat at least three times to remove trypsin (if it was used in step 1 above) and all remaining media, which in the case of serum-supplemented cell-culture media may contain very large amounts of albumin and other proteins.

3. Resuspend and pellet the cells once more in wash buffer.
4. Lyse the cells with a volume of cell lysis buffer such that the final protein concentration should be 5–10 mg/mL (*see* **Note 8**).
5. Centrifuge the cell lysates and confirm with pH paper that the pH of the supernatant is still at 8.5.
6. If the pH is not correct, re-check the cell wash and lysis buffer pH and repeat the process, or adjust the pH by careful addition of 50 m*M* HCl or NaOH (*see* **Note 9**). At this juncture, a decision should be made whether to perform a sample precipitation to further purify the sample (*see* **Subheading 3.2.** and **Note 1**).
7. Store cell lysates in aliquots at –80°C while protein concentration assays are carried out on aliquots of all lysates (*see* **Note 10**).
8. Use hydrolysis/amino acid analysis or the 2-D Quant Kit (GE/Amersham), to determine the protein concentrations in all samples.
9. Use 1.5-mL polypropylene microcentrifuge tubes placed on ice to aliquot each sample. Adjust the volume of each sample to as close to 50 μg/10 μL as possible. Normalize the volume of all the samples within a given experiment.
10. In vitro label each of the control and experimental protein extracts with Cy3 and Cy5 dyes respectively. If a pooled internal standard will be used (i.e., as an aid to comparing and normalizing multiple experiments), label it with Cy2. For labeling each sample use 1 μL of each dye solution that has been diluted from stock to 400 pmol/μL with fresh dimethyl formamide (*see* **Note 11**).
11. Incubate each reaction for 30 min at 4°C.
12. Add 1 μL of 10 m*M* lysine to each reaction tube. Incubate for an additional 10 min on ice to stop each labeling reaction.
13. Working quickly, pool the control, experimental, and internal standard (if used) samples together (i.e., 150 μg total protein) into onc of the sample tubes. Wash the individual sample tubes with 50 μL rehydration buffer, microfuge the tubes, and combine all traces of the sample tube washes into one of the pooled sample tube. Dilute the pooled sample with additional rehydration buffer to a final volume of between 400 and 450 μL (*see* **Note 12**). If necessary, the pooled sample may be frozen and stored at this point at –80°C; however, it is best to proceed directly with the IEF.

3.1.2. Labeling With Saturation CyDyes

The sample preparation for labeling with saturation CyDyes is essentially the same as that for minimal dyes, except that the lysis/labeling buffer is pH 8.0 at 37°C. The pH difference is critical. The subsequent labeling procedure, however, varies substantially.

For analytical (5 μg/sample) labeling with saturation CyDyes:

1. Using a clean, sterile 1.5-mL polypropylene microfuge tube for each dye/sample pair, dispense 5 μg protein in 9 μL lysis/labeling buffer.
2. Add the required volume of 2 m*M* Tris(2-carboxyethyl)phosphine hydrochloride (TCEP) (usually approx 1 μL, *see* **Note 13**).

3. Mix thoroughly by pipetting and spin briefly. Incubate for 1 h at 37°C in the dark.
4. Add the required volume of 2 m*M* dye in dimethylformamide (DMF) (usually approx 2 μL *see* **Note 13**).
5. Mix thoroughly by pipetting and spin briefly. Incubate for 30 min at 37°C in the dark.
6. Stop the reaction by adding an equal volume of 2X sample buffer. Mix thoroughly by pipetting and spin briefly. Store at –70°C in the dark or further dilute in resuspension buffer and proceed with the IEF run.

For preparative labeling of a sample of interest using Cy3 saturation CyDye (*see* **Note 14**):

1. Dispense 500 μg of the protein sample of interest (may be a pool of samples, if necessary) in lysis/labeling buffer into a clean, sterile 1.5-mL polypropylene microfuge tube.
2. Add 10 μL TCEP at the required concentration (*see* **Note 13**).
3. Mix thoroughly by pipetting and spin briefly. Incubate for 1 h at 37°C in the dark.
4. Add the required volume of 20 m*M* Cy3 preparative CyDye in DMF (*see* **Note 13**).
5. Mix thoroughly by pipetting and spin briefly. Incubate for 30 min at 37°C in the dark. Add resuspension buffer to a final volume between 400 and 450μL (assuming the use of a 24-cm IPG strip). Store at –70°C in the dark or (preferably) proceed with the IEF run.

3.2. Sample Precipitation Using Organic Solvent/TCA or the GE/Amersham 2D Clean-Up Kit

This procedure is optional, although it is generally recommended; *see* **Note 1**.

3.3. Isoelectric Focusing: First-Dimension Separation

3.3.1. Rehydration Loading

1. Transfer the pooled sample into the IPG strip holder, spreading the sample evenly across the length of the holder between the electrodes while avoiding air bubbles.
2. Remove the protective plastic backing from the IPG strip of choice (*see* **Note 15**) and, starting with one end, carefully place it gel side down onto the solution, being careful not to trap air bubbles or to abrade/damage the gel. If necessary, the gel may be gently lifted to reposition it for proper electrode contact, and trapped bubbles may be removed by very gently tapping the backing of the IPG strip with a (clean) pipet tip.
3. Overlay the strip with 3 mL of Dry Strip mineral oil. Place the holder onto the IEF apparatus.
4. Suggested running conditions for rehydration loading are as follows, all at 20°C:

2 h passive re-hydration	with	no applied voltage
12 h	at	30 V
1 h	at	500 V
8 h	at	8000 V

After the run is complete, the voltage is ramped down to 500 V to maintain focusing until the gel strip is removed.

5. Remove the mineral-oil overlay by pouring or pipetting it off the gel strip. Carefully place the IPG strip with the gel side up into an equilibration tube. If necessary, after IEF, the IPG strip may be stored in an equilibration tube at –70°C, but it is best to proceed directly to the next steps.
6. Place 10 mL of equilibration buffer containing DTT into the tubes containing the IPG strips, incubate all tubes for 10 min with gentle mixing on a rocking platform.
7. Completely remove the DTT-containing solution and replace it with equilibration solution containing iodoacetamide. Incubate for an additional 10 min. (Important: skip this step for saturation dye labeling.)
8. Pour off the equilibration solution and briefly rinse the strip with 15 mL of 1X gel running buffer.

3.3.2. Cup Loading for Analytical Saturation CyDye Gels (for the GE/Amersham System)

3.3.2.1. Preparation of the IPG Drystrip: Rehydration for Cup Loading (Analytical Gels)

1. Pipet the appropriate volume of rehydration buffer into each of the required number of slots in an Immobiline DryStrip Reswelling Tray, or equivalent vessel. The volume should not exceed 450 μL per 24-cm strip.
2. Deliver the buffer slowly along the slot. Remove any large bubbles.
3. Remove the protective cover from the Immobiline DryStrip (IPG strip).
4. Position the IPG strip with the gel side down and the pointed (acidic) end of the strip against the end of the slot closest to the spirit level.
5. Lower the IPG strip onto the buffer. To help coat the entire IPG strip and avoid air bubbles, gently lift and lower the strip along the surface of the buffer.
6. Overlay each IPG strip with 1.5–2 mL DryStrip (mineral oil) to prevent evaporation and urea crystallization.
7. Cover the DryStrip Reswelling Tray and allow the IPG strip to rehydrate at room temperature. Rehydrate 16 h (overnight), but not longer than 24 h.

3.3.2.2. Cup Loading

1. Pre-prepare electrode pads by cutting 5 mm × 15 mm pieces from the IEF Electrode Strips. Place the pieces on a glass plate and soak with distilled water. Remove excess water by blotting with a paper towel, or filter paper. It is important that the pads are damp but not wet, as excess water may cause streaking.
2. Combine the required amount of each labeled protein extract (e.g., 5 μg Cy3 pooled internal standard, 5 μg Cy5 experimental sample).
3. Mix thoroughly by pipetting and leave on ice until use.
4. Place the Ettan IPGphor Cup Loading Strip Holders in the correct position on the IPGphor platform.
5. With a pair of forceps, carefully remove the IPG strip from the DryStrip Reswelling Tray, taking care not to damage the gel.

6. Allow the excess DryStrip mineral oil to run off the IPG strip onto a piece of tissue. Do not allow the gel side to touch the tissue as it may stick to it.
7. Place the IPG strip gel side up with the basic end (flat end of the IPG strip) flush with the flat end of the IPGphor Cup Loading Strip Holder.
8. Place preprepared damp electrode pads (5 mm × 15 mm) onto the acidic and basic ends of the gel.
9. Clip down the electrodes firmly onto the electrode pads. Ensure that there is good contact with the IPG strip and the metal on the outside of the strip holder.
10. Clip a cup loader onto the strip next to one of the electrodes. It should be positioned either at the acidic or basic end (*see* **Note 16**), in between the two electrodes.
11. Check for a good seal by filling the cup to the top with DryStrip mineral oil. Observe the level of the fluid to determine whether it is decreasing. If a leak is detected, remove the mineral oil and reposition the sample cup.
12. Apply at least 4 mL of DryStrip mineral oil onto the DryStrip holder. Allow the oil to spread so it completely covers the IPG strip.
13. Up to 100 μL of protein sample can now be loaded into the bottom of the sample cup.
14. Put the clear plastic strip cover onto the strip holder.
15. Cover the apparatus to exclude light while ensuring that the air vents remain uncovered. Metallic covers must not be used under any circumstances. The IPG strips are now ready for IEF.
16. Focus the proteins overnight. Suggested room temperature running conditions are given below for analytical analyses with cup loading onto 24 cm, pH 3.0–10.0 strips.

3 h	at	300 V
3 h		Linear gradient from 300 to 600 V
3 h		Linear gradient from 600 to 1000 V
3 h		Linear gradient from 1000 to 8000 V
4 h	at	8000 V
hold	at	500 V
		50 μA upper current limit per strip

17. Strips should be removed as soon as possible after the end of the 4 h/8000 V hold step is completed. If they are left for more than 2 h at 500 V, strips should be ramped up linearly to 8000 V over 30 min to refocus proteins before the strips are removed from the apparatus.

3.4. SDS-PAGE, Second-Dimension Separation

3.4.1. Preparing the Plates

Reflective reference markers are used by the gel-picking software to determine the spot coordinates. Gels for spot picking must therefore be cast with two reference markers under the gel, and the gel has to be bound to the glass plate to ensure that it does not deform during the picking process.

1. Treat the shorter low-fluorescence glass plate with diluted Bind-Silane solution. Pipet 2 mL of Bind-Silane solution over the surface of the plate and wipe with a lint-free tissue until dry. Cover the plate with a lint-free tissue and leave on the bench or in a very clean hood for 1.5 h to allow the excess Bind-Silane to evaporate.
2. Separately treat the larger plate with undiluted Repel-Silane, as in step 1. Be sure to let the plates dry in an area that is well separated from the Bind-Silane plates. Otherwise, the vapors from each can cross-contaminate the other mating plate.
3. Once dry, any residue on the glass plates may be removed by cleaning with a lint-free tissue that has been slightly dampened with ethanol. Place a reference marker halfway down the left side of the Bind-Silane treated plate, close to but not touching the spacer, in a position where it will not likely interfere with expected protein spot(s). With a second marker, repeat this procedure on the right side of the plate.
4. When ready to pour the gel, sandwich the Bind-Silane-treated plate against the Repel-Silane-treated glass plate. Place the glass plate cassette in the gel caster. Ensure that the plates have completely dried. It is best to assemble the silanized plates immediately prior to gel pouring to avoid transfer of volatile Bind-Silane between glass surfaces within the cassette.

3.4.2. Preparing the Gel Solution and Casting Gels for Isocratic SDS-PAGE

1. Allow the acrylamide gel stock solution to warm to room temperature. It may be sonicated or filtered through a 0.2-μm filter to de-gas.
2. Assemble the gel caster on a level surface.
3. Connect one end of the feed tube to either a funnel or a peristaltic pump. Insert the opposite end into the grommet in the bottom of the balance chamber.
4. Pour 100 mL of displacing solution into the balance chamber.
5. When ready to pour the gels, add the appropriate volume of freshly prepared APS and TEMED to the acrylamide gel stock solution and mix.
6. Introduce the gel solution into the funnel or peristaltic pump, taking care not to admit any air bubbles into the feed tube.
7. Allow the solution to enter the gel caster until it is 1–2 cm below the final desired height. Stop the flow of acrylamide and remove the feed tube from the balance chamber grommet. Once the feed tube is removed, the dense displacing solution will enter the caster and force the remaining acrylamide solution into the gel cassettes to the desired height.
8. Immediately pipet 1–2 mL of water-saturated butanol onto each gel to create a level interface.
9. Allow the gels to polymerize for at least 3 h at room temperature before disassembling the caster. The gels can be stored in an airtight container at 2–8°C covered with 1X SDS electrophoresis running buffer for up to 2 wk.

3.4.3. SDS-PAGE Procedure

1. Heat sufficient low-melting agarose (dissolved in 1X sample buffer containing a trace amount of BPB) in a microwave until it reaches near boiling. Fill the void at the top of the gel cassette sandwich with 1X buffer to facilitate placing the IPG gel strip onto the top of the gel. This serves to reduce the friction when positioning the strip onto the top of the gel and to break the surface tension at the top of the gel, which aids in removing trapped bubbles. Place the strip gently onto the top of the gel, noting the orientation of the strip. Quickly remove trapped air bubbles between the strip and the top of the gel by gently tapping the upper edge of the IPG strip with a thin piece of plastic, being careful not to contact and damage the IPG gel itself. Ensure that the strip is in direct contact with the polyacrylamide gel.
2. Pour off the 1X sample buffer (to prevent SDS and buffer from diffusing out of the gel do not let the gel stand in the sample buffer longer than necessary) and immediately overlay the gel with the melted agarose solution. Remove any remaining trapped bubbles immediately after pouring. Again, ensure that the strip is in direct contact with the polyacrylamide gel.
3. Allow the gel to cool before placing it into the vertical gel apparatus, making sure that the agarose has solidified.
4. Run the gel at 15°C. Use approx 20 mA current per gel for an 8-h run, or approx 10 mA current for a 16-h, overnight run.

3.5. Sample Imaging

1. Immediately after SDS PAGE, the gel (which is still held between two glass plates) is sequentially scanned at 2 (for two dyes) or 3 (if a multiplexing experiment is performed with an internal standard) wavelengths on an Amersham Typhoon 9410 Imager. After scanning, 16-bit gel files of each color channel are exported for image analysis using the differential in-gel analysis module of the Amersham DeCyder software package. After spot detection (which includes automatic background correction, spot volume normalization, and calculation of the volume ratio), a user-defined filter may be applied to each gel. This has the effect of automatically removing many nonprotein spot features from the gel and is followed by recalculation of sorting and normalization parameters.
2. The front glass plate is removed and the gel is then fixed and stained with Sypro Ruby or Deep Purple, fluorescent stains that will be used as a guide to excise spots of interest from the gel (*see* **Notes 17** and **18**).
3. To fix the gels prior to Sypro Ruby staining, the gel is placed in 200 mL of 50% methanol, 10% acetic acid fixative for 2 h to overnight at room temperature with gentle shaking. Pour off the fixative solution.
4. Rinse the gel for 4 × 15 min in 400 mL water; pour off the rinse water.
5. Incubate the gel in Sypro Ruby stain for 5 h to overnight. Pour off and properly dispose of the stain solution.
6. Rinse the gel briefly with 400 mL water; pour off the rinse water.
7. De-stain with 10% methanol/6% acetic acid for 0.5 h.

8. Rinse the gel 2 × 15 min with 300 mL water.
9. Replace the top plate, taking care to remove all bubbles, and re-scan using the appropriate wavelength.
10. Amersham Biosciences DeCyder software is then used to quantify the gel image and to identify a pick list of differentially expressed protein spots to be excised and subjected to in gel trypsin digestion followed by MS-based protein identification. The DeCyder software can analyze any two CyDye gel images, either on the same gel or on different gels, match the spots between the two images, and then identify differentially expressed protein spots. The DeCyder software automatically outputs a listing of statistically significant differences in protein expression including *t*-test values, using the Cy2 internal standard. Replicate samples are required for statistical analysis. Differentially expressed spots may be identified using a number of possible criteria, including peak area, peak volume, three-dimensional (3D) peak slope, 3D peak height, and/or statistical parameters. Protein spots that show significant differences in their intensities between the two samples will be highlighted by the software so they can be manually confirmed. **Figure 1** shows an example of a DIGE analysis of a complex protein extract from cells grown in culture. This gel has been scanned to allow visualization of the Cy3-labeled control (**Fig. 1A**), the Cy5-labeled experimental sample (**Fig. 1B**) and a Cy3-labeled control (**Fig. 1C**), which shows those spots that are differentially regulated by threeold or more based on spot volume ratios in the control vs experimental samples. The DeCyder software can also analyze Sypro Ruby images, match the spots found with Sypro staining to those identified with the CyDye stains, and then choose a pick list from the Sypro-stained gel image. DeCyder data can be read by laboratories who do not have the DeCyder software by exporting the data using an HTML format.

3.6. Spot Picking, Tryptic Digestion

1. The protein spot pick list is transferred to the Ettan Spot Picker instrument (Amersham Biosciences/GE), which automatically excises the selected protein spots (using a 2.0 mm diameter pick head) from the gel and transfers them into a polypropylene 96-well microtiter plate (Corning, Corning, NY) (*see* **Note 19**).
2. The excised protein spots are then subjected to automated in-gel tryptic digestion on the Ettan TA Digester, or manually digested using the following steps:
    a. Wash the spots with 2 × 50 µL of 50 m*M* ammonium bicarbonate in 50% CH_3CN for 10 min.
    b. Wash the spots with 1 × 50 µL 75% CH_3CN for 10 min.
    c. Air-dry for 15–30 min.
    d. Add 15 µL (0.1 µg) trypsin per spot; add proportionally more if a significantly larger gel piece is manually excised.
    e. Incubate for 3 h to overnight at 37°C.

LC-MS/MS or matrix-assisted laser desorption/ionization time-of-flight MALDI-TOF)/TOF MS/MS provides sensitive and accurate mass spectral data

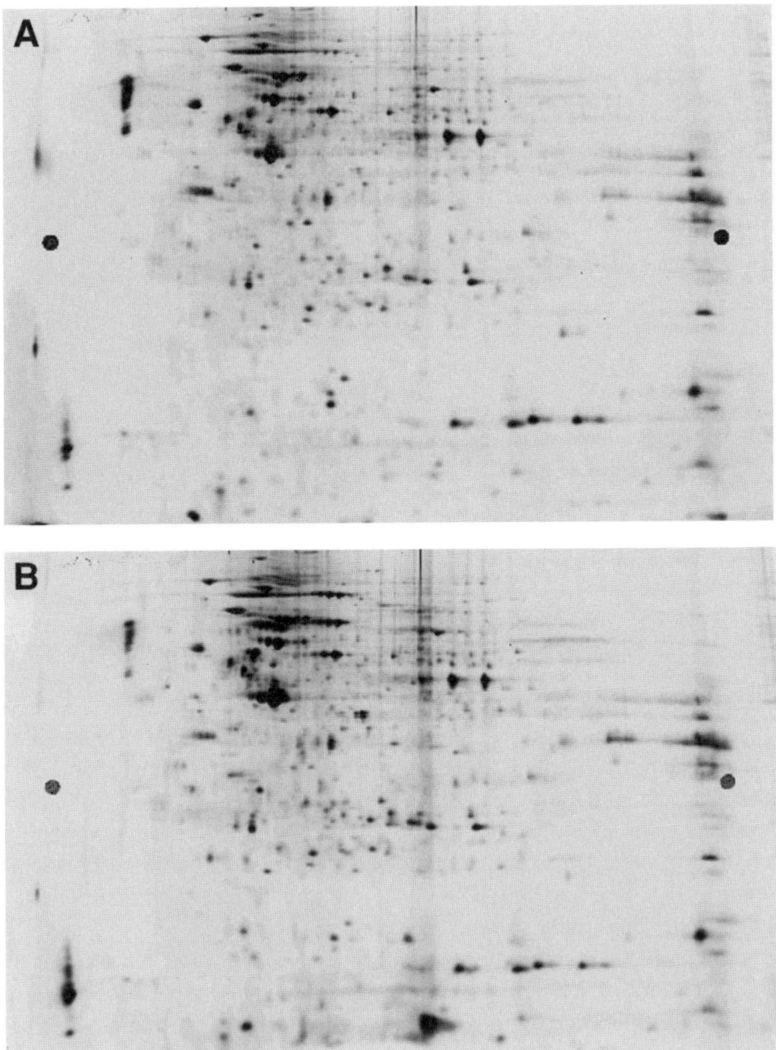

Fig. 1. **(A)** Control Cy3-labeled sample. **(B)** Experimental (treated) Cy5-labeled sample. **(C)** Spots outlined display threefold or greater changes in spot volume ratio. Increases in Cy3/Cy5 ratio are indicated by (*), increases in Cy5/Cy3 ratio are unmarked. *(figure continues)*

on the in-gel tryptic digests. The resulting, uninterpreted MS/MS spectra may then be subjected to database searching using Mascot and other commercially available algorithms to enable high-throughput protein identification.

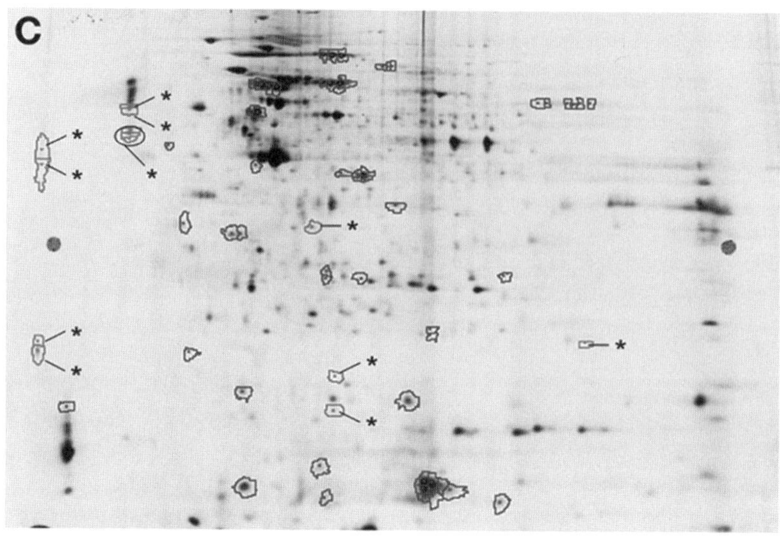

Fig. 1. *(continued)*

4. Notes

1. Precipitation of the protein sample is an effective method for separating and removing many contaminants, both chemical (such as salts, ionic detergents, buffers, and other small molecules) and biological (such as endogenous small ionic molecules, nucleotides, phospholipids, lipids, polysaccharides), leaving behind a much purer protein fraction that can then be resuspended in a buffer ideally suited for both CyDye labeling and 2D gel electrophoresis. The method of choice for precipitation may be largely sample dependent.
   a. Organic solvent precipitation (acetone, acetonitrile or ethanol): add three volumes or more of ice-cold solvent to the aqueous extract; allow the proteins to precipitate at –20°C for at least 2 h. The proteins are pelleted at high force (12,000*g*) and residual solvent is dried off to near completion before resuspension in the labeling/lysis buffer.
   b. Organic solvent (acetone, acetonitrile, or ethanol) with trichloroacetic acid (TCA): although this is more effective than solvent or TCA alone, extended exposure to low pH may cause protein degradation or modification. Use three or more volumes of solvent containing 10 to 20% TCA with 20 m*M* DTT or TCEP, and precipitate for at least 2 h at –20°C. Pellet the proteins by centrifugation as in **step a**, and wash the pellet in cold solvent. Remove solvent to near dryness before resuspension in lysis/labeling buffer.
   c. 2D clean-up kit (GE/Amersham; *see* **Subheading 2.2.**): this kit has been demonstrated in several systems to be efficient, with high and reproducible recovery, and is highly recommended for general use.

1. When using precipitation or any other fractionation method, partial sample loss and reproducibility between sample pairs or groups is always a concern, and must be empirically accessed on a sample-by-sample basis. However, protein precipitation is often a necessary and worthwhile risk if it results in more clearly defined gel patterns and minimization of streaking and other gel artifacts.
2. The addition of ampholytes in the lysis buffer is a common practice in the preparation of protein samples for analysis in traditional 2D gels, as they enhance protein solubility by minimizing charge–charge induced interactions, but would severely decrease the labeling efficiency of minimal labeling CyDyes in a DIGE 2D experiment. Conversely, ionic detergents, such as SDS, are often used during protein extraction and/or solubilization, and although the CyDye labeling process is largely unaffected by the presence of SDS (at levels up to at least 1%), the presence of SDS can strongly interfere with the subsequent running of the first-dimension IEF. The negatively charged SDS–protein complex will not focus properly unless the SDS is removed or sequestered. In some cases, however, the inclusion of SDS may be required. In these instances the deleterious effects of SDS on the first-dimension IEF run can be minimized by diluting the SDS-containing sample into a rehydration solution containing a zwitterionic or non-ionic detergent (e.g., CHAPS, up to 4% or higher; NP-40, up to 1%; ASB-14, up to 2%; or Triton X-100, up to 0.5%) so that the final concentration of SDS is 0.25% or lower and the ratio of the other detergent(s) to SDS is at least 8:1 (w:w) ***(25)***. Alternatively, if the SDS is a contaminant, it may be at least partially removed by precipitation of the protein. Some detergents, such as ASB-14 and Triton X-114, have been shown to significantly enhance the solubility of membrane/hydrophobic proteins ***(25–30)***. The mixture of thiourea with urea has also been reported to increase the solubility of membrane proteins ***(31)***. *See also* the GE/Amersham CyDye product notes (25800983/RPK0272PL) ***(24)*** for additional and updated information about dye compatibility and protocols.
3. Nucleic acid (NA) contamination is often problematic, especially in the preparation of high protein concentration samples (>5 mg/mL protein), in which a minimal volume of lysis buffer is used to solubilize collected cell or tissue samples, and a coincidentally high content of NA remains in solution with the extracted proteins. The consequences of NA contamination can be significant, and include nonhomogenous samples, which are difficult or impossible to accurately aliquot, leading to unreliable results when comparing one sample to another or when running technical replicates of aliquots of the same sample. Nucleic acids can bind proteins through electrostatic interactions, thus preventing and skewing accurate and reproducible sampling. In addition, high-molecular-weight NA can severely interfere with proper gel running and can cause distortion and smearing in the final 2D gel. Finally, it is not clear how the presence of high-molecular-weight NA may affect sample labeling with CyDyes. In samples where NA contamination is a problem, they may be effectively removed by treatment of the sample with a (protease-free) DNase/RNase such as Benzonase (EMD Chemicals, San Diego, CA). Addition of a solution containing 1 mg/mL Benzonase and 50 m*M* $MgCl_2$ followed by incubation on ice is sufficient to reduce the NA to mono- and oligonucleotides. These can subsequently be removed by precipitation of the protein fraction.

4. Reduction of a sample's complexity before CyDye labeling may be necessary in order to label and visualize lower-abundance proteins, as some samples may contain tens of thousands of proteins that span a larger dynamic range in concentrations than can be approached with DIGE or other proteomic technologies. Prefractionation may be performed in a number of effective ways at several different levels, including preparative isoelectric focusing, chromatographic techniques, protein extraction, organellar purification, or laser capture microdissection (LCM) techniques. Differential sample loss between experimental and control groups is always a concern with any fractionation steps. Alternative approaches that have been used to reduce the complexity of serum-derived or -associated samples are depletion technologies that may, for instance, utilize bead-bound specific antibodies or other reactive groups that are targeted towards the removal of specific, high-abundance species such as albumin and immunoglobulins. These proteins account for the majority of the protein mass in serum samples, and thus may interfere with the labeling of low-abundance species, as the dyes label by the law of mass action and will therefore be proportionally sequestered by the most abundant proteins present in the mixture. Many depletion kits are currently available; some (Genway, Gibbstown, NJ) can be tailored to selectively remove 12 or more components on a single column pass. In addition to preventing labeling of less-abundant proteins, highly abundant proteins also will prevent loading a sufficient total amount of protein onto gels to permit the detection of less-abundant proteins.
5. The same buffer is used for both lysis and labeling, unless conditions dictate that a different buffer or condition must be used for lysis, in which case the sample should be precipitated and then re-suspended in the standard labeling buffer before attempting to label and run the sample.
6. Once dissolved in urea, the protein samples should never be heated, because considerable charge heterogeneity can be introduced as a result of carbamylation of the proteins by isocyanate, which is formed during the decomposition of urea.
7. Trypsinization would not be appropriate if cell-surface proteins are of interest. In general, trypsinization should be avoided in sample preparations targeted for proteomic analysis. If it is used, additional pellet resuspensions and washings may be required to completely remove trypsin prior to cell lysis.
8. 1×10^6 tissue-culture cells contain approx 50 µg protein, so the volume of lysis buffer used should be proportioned accordingly to arrive at a final concentration of approx 5 mg/mL total protein. Samples may be sonicated with several brief low- to medium-power pulses, while kept on ice. Although this effectively solubilizes difficult samples, there is a danger of protein degradation, particularly if the samples are allowed to overheat during sonication, and differential degradation is a possible complication. Also at this juncture, it should be determined whether it is necessary to nuclease treat the lysed cell suspension and/or precipitate the proteins. In the latter case, the precipitated protein should be resuspended in a minimum volume of lysis buffer prior to proceeding to **step 8**.

9. If the pH of the sample is below 8.0 for minimal labeling, or not within 0.2 U of 8.0 for saturation labeling, then it must be adjusted before proceeding, as the labeling efficiency falls off significantly if pH is outside of these ranges. An alternative to the slow addition of 50 m*M* NaOH or HCl is to make up lysis buffer (7 *M* urea, 2 *M* thiourea, 4% CHAPS, 30 m*M* Tris at the appropriate temperature) using Tris at pH 9.5 or 7.5, as needed. It is also important not to over-dilute the sample, as this will also affect the labeling efficiency.
10. If the samples are to be stored frozen for later use or for repeat experiments, tests should be carried out to determine whether freeze/thawing leads to overall or differential protein loss from precipitation.
11. The pooled internal standard consists of equal proportions of each sample in the experiment (control + experimental) combined to give a total of 50 μg pooled protein sample/gel (i.e., 50 μg of Cy3-labeled control sample, 50 μg Cy5-labeled experimental sample, and 50 μg Cy2 labeled pooled sample are combined to be run on a single gel. Labels for the control and experimental samples may be interchanged. Large sample sets will require staggered timing.
12. An unlabeled aliquot of protein containing equal amounts of both the control and experimental protein sample may be added to the sample mixture (not to exceed approx 1 mg total protein loaded on the gel) to increase the probability of identifying picked spots by the subsequent mass spectroscopic method employed for analysis. The amount of protein that can be loaded must be optimized for each individual experiment, as overloading sample will distort the protein patterns, particularly in the vicinity of higher-abundance species, as well as affect the overall running and quality of the gels. The advantage of using additional unlabeled as opposed to labeled protein is the cost savings from having to use less dye.
13. The molar TCEP:dye ratio should be maintained at 1:2 to ensure efficient labeling. Samples with higher cysteine content will require more TCEP to reduce the disulfide bonds, and more dye to label the thiol groups. Typically, 5 μg of protein lysate requires 2 nmol TCEP and 4 nmol dye for the labeling reaction (assuming an average cysteine content of 2%). The criteria for optimal labeling conditions are as follows:

    a. All spots must overlay;
    b. There must be no significant mass "trains" or vertical streaks;
    c. There must be no significant charge "trains" or horizontal streaks.

    These criteria may be assessed by overlaying the resulting Cy3 and Cy5 images using the image capture software associated with the scanner, e.g., ImageQuant(tm) (Molecular Dynamics).

    If the amount of TCEP/dye is too low, available thiol groups on some proteins will not be labeled. When the maleimide dye labels a thiol group, the mass of the protein is increased but the charge is unaffected. Under-labeled samples will show MW trains and/or streaking in the vertical direction. Also, when the amount of TCEP/dye is too low, differential migration of Cy3- and Cy5-labeled spots for the same protein can also occur.

If the amount of TCEP/dye is too high, this can lead to nonspecific labeling of the amine groups on protein lysine residues. When the maleimide dye labels a lysine group, the mass of the protein is increased and the charge is also reduced by 1. Over-labeled samples will show pI charge trains and/or streaking in the horizontal direction (see GE/Amersham guide for CyDye DIGE Fluor Labeling kit for scarce samples, 25800983PL, for more detail).

14. For the saturation dye preparative gels, it is best to target approx 500 µg, but no more than 1 mg, of either the pooled sample or the sample with the highest-abundance spot of interest for labeling. GE/Amersham manufactures a Cy3 saturation dye specifically for the purpose of preparative gel labeling (kit no. 25-8009-84); alternatively, Sypro Ruby staining may be performed instead.
15. IPG strips are available with a wide variety of pH ranges; GE/Amersham manufactures 24-cm Immobiline IPG Drystrips in pH 3.0–10.0 and 3.0–11.0 broad ranges. Other pH ranges that are available include: 3.0-5.6 NL, 3.0–7.0 NL, 4.0–7.0, 3.5–4.5, 4.0–5.0, 4.5–5.5, 5.0–6.0, 5.3–6.5, 5.5–6.7, and 6.0–9.0. As noted, some ranges are available in both linear and nonlinear (NL) formats.
16. Cathodic cup loading may give better results for acidic IPG strips. Higher protein loads (e.g., for preparative gels) may require longer focusing times. *See* GE/Amersham guide 25800983PL for more detail about cup loading.
17. As a result of the lag time between scans, for large experiments involving multiple gels it will be necessary to first fix the gels before scanning to avoid sample diffusion over time.
18. The reason for poststaining all of the protein in the gel is so that the CyDye labeling is carried out such that the extent of incorporation will be less than 5% in terms of mole CyDye/mole protein. Because the CyDye adds about 460 Da to each protein, low-MW proteins (e.g., <10 kD) labeled with CyDyes will not precisely co-migrate in the SDS-PAGE dimension with their nonlabeled counterparts. Thus, when picking the spots from the gel with a spot-picking robot, the central mass of the spot may be missed, and consequently less material will be available for protein identification. After poststaining, the gel is rescanned and picking coordinates are readjusted accordingly. This poststaining is not an issue with saturation CyDyes, which label every cysteine of every protein molecule in solution. Protein spots can be picked directly from gels that were run with proteins labeled with CyDye saturation dyes.
19. The maximum spot-pick head on the Ettan Spot Picker is 2.0 mm. Many spots are larger than this and, as a result, much of the protein in these spots may be left behind in the gel. The spot picker may be programmed to pick multiple plugs to try to excise as much of the spot as possible. Alternatively, the spot may be carefully excised manually with a sharp tweezer or similar tool. A post-spot-picking scan may be performed in order to assess the completeness of excision of spots of interest.

Acknowledgments

The author thanks Dr. Sherman Weissman for providing samples and for granting permission to use the figures in this chapter. Also, thanks to Dr. Kenneth Williams for critical reading of the manuscript and for insightful discussions. This project has been funded in part with federal funds from National Heart, Lung and Blood Institute/National Institutes of Health (NIH) contract N01-HV-28186 and National Institute on Drug Abuse/NIH grant 1 P30 DA018343-01.

References

1. Görg, A., Weiss, W., and Dunn, M. (2004) Current two-dimensional polyacrylamide gel electrophoresis technology for proteomics *Proteomics* **4,** 3665-3685.
2. Alban, A., David, S. O., Bjorkesten , L., et al. (2003) A novel experimental design for comparative two-dimensional gel analysis: Two-dimensional difference gel electrophoresis incorporating a pooled internal standard. *Proteomics* **3,** 36-44.
3. Ünlü, M., Morgan, M. E., and Minden, J. S. (1997) Difference gel electrophoresis: A single gel method for detecting changes in protein extracts. *Electrophoresis* **18,** 2071-2077.
4. Tonge, R., Shaw, J., Middleton, B., et al. (2001) Validation and development of fluorescence two-dimensional differential gel electrophoresis proteomics technology. *Proteomics* **1,** 377-396.
5. Zhang, H., Yan, W., and Aebersold, R.(2004) Chemical probes and tandem mass spectrometry: a strategy for the quantitative analysis of proteomes and subproteomes. *Curr. Opin. Chem. Biol.* **8,** 66-75.
6. White, I. R., Pickford, R., Wood, J., Skehel, J. M., Gangaharan, B., and Cutler, P. (2004) A statistical comparison of silver and SYPRO Ruby staining for proteomic analysis. *Electrophoresis* **25,** 3048-3054.
7. Shaw, J., Rowlinson, R., Nickson, J., et al. (2003) Evaluation of saturation labelling two-dimensional difference gel electrophoresis fluorescent dyes. *Proteomics* **3(7),** 1181-1195.
8. Anderson, N. L. and Anderson, N. G. (2003) The human plasma proteome: history, character, and diagnostic prospects. *Mol. Cell. Proteomics* **2,** 50.
9. Corthals, G. L., Wasinger, V. C., Hochstrasser, D. F., and Sanchez, J. (2000) The dynamic range of protein expression: A challenge for proteomic research. *Electrophoresis* **21,** 1104-1115.
10. Van Den Bergh, G., Clerens, S., Vandesande, F., and Arckens, L. (2003) Fluorescent two-dimensional difference gel electrophoresis and mass spectrometry identify age-related protein expression differences for the primary visual cortex of kitten and adult cat. *J. Neurochem.* **85(1),** 193-205.
11. Hu, Y., Wang, G., Chen, G. Y., Fu, X., and Yao S. Q. (2003) Proteome analysis of *Saccharomyces cerevisiae* under metal stress by two-dimensional differential gel electrophoresis. *Electrophoresis* **24,** 1458-1470.
12. Macdonald, N., Chevalier, S., Tonge, R., et al. (2001) Quantitative proteomic analysis of mouse liver response to the peroxisome proliferator diethylhexylphthalate (DEHP) *Arch. Toxicol.* **7,** 415-424.
13. Ruepp, S. U., Tonge, R. P., Shaw, J., Wallis, N., and Pognan, F. (2002) Genomics and proteomics analysis of acetaminophen toxicity in mouse liver. *Toxicol. Sci.* **1,** 135-150.

14. Yan, J. X., Devenish, A. T., Wait, R., Stone, T., Lewis, S., and Fowler, S. (2002) Fluorescence two-dimensional difference gel electrophoresis and mass spectrometry based proteomic analysis of *Escherichia coli*. *Proteomics* **2,** 1682-1698.
15. Knowles, M. R., Cervino, S., Skynner, H. A., et al., (2003) Multiplex proteomic analysis by two-dimensional in-gel electrophoresis. *Proteomics* **3,** 1162-1171.
16. Swatton, J. E., Prabakaran, S., Karp, N. A., Lilley, K. S., and Bahn, S. (2004) Protein profiling of human postmortem brain using 2-dimensional fluorescence difference gel electrophoresis (2-D DIGE). *Mol. Psychiatry* **2,** 128-143.
17. Kernec, F., Unlu, M., Labeikovsky, W., Minden, J. S., and Koretsky, A. P. (2001) Changes in the mitochondrial proteome from mouse hearts deficient in creatine kinase. *Physiol. Genomics* **6,** 117-128.
18. Hoffert, J. D., van Balkom, B. W., Chou, C. L., and Knepper, M. A.(2004) Application of difference gel electrophoresis to the identification of inner medullary collecting duct proteins. *Am. J. Physiol. Renal Physiol.* **286(1),** F170-F179.
19. Vierstraete, E., Verleyen, P., Baggerman, G., et al. (2004) proteomic approach for the analysis of instantly released wound and immune proteins in Drosophila melanogaster hemolymph. *Proc. Natl. Acad. Sci. USA* **101(2),** 470-475.
20. Gharbi, S., Gaffney, P., Yang, A., et al. (2002) Evaluation of two-dimensional differential gel electrophoresis for proteomic expression analysis of a model breast cancer system. *Mol. Cell. Proteomics* **1,** 91-98.
21. Zhou, G., Li, H., Decamp, D., et al. (2002) 2D Differential In-gel Electrophoresis for the identification of esophageal scans cell cancer-specific protein markers. *Mol. Cell. Proteomics* **1,** 117-124.
22. Lee, J. R., Baxter, T. M., Yamguchi, H., et al. (2003) Differential protein analysis of spasomolytic polypeptide expressing metaplasia using laser capture microdissection and two-dimensional difference gel electrophoresis. *Appl. Immunohistochem. Mol. Morphol.* **11,** 188-193.
23. Friedman, D. B., Hill, S., Keller, J. W., et al. (2004) Proteome analysis of human colon cancer by two-dimensional difference gel electrophoresis and mass spectrometry. *Proteomics* **3,**793-811.
24. Ettan DIGE Fluors User Manual, Amersham Biosciences (2004) RPK0272PL (minimal Dyes) and 25800983PL (saturation dyes).
25. Ames, G. F. (1976) Two-dimensional gel electrophoresis of membrane proteins. *Biochemistry* **15,** 616-623.
26. Santoni, V., Molloy, M., and Rabilloud, T. (2000) Membrane proteins and proteomics: un amour impossible? *Electrophoresis* **21,** 1054-1070.
27. Rabilloud, T., Blisnick, T., Heller, M., et al. (1999) Analysis of membrane proteins by two-dimensional electrophoresis: Comparison of the proteins extracted from normal or *Plasmodium falciparum*-infected erythrocyte ghosts. *Electrophoresis* **20,** 3603 -3610.
28. Chevallet, M., Santoni, V., Poinas, A., et al. (1998) New zwitterionic detergents improve the analysis of membrane proteins by two-dimensional electrophoresis. *Electrophoresis* **19,** 1901 -1909.

29. Pierce, M. J., Wait, R., Begum, S., Saklatvala, J., and Cope, A. P. (2004) Expression profiling of lymphocyte plasma membrane proteins. *Mol. Cell. Proteomics* **3,** 56-65.
30. Luche, S., Santoni, V., and Rabilloud, T. (2003) Evaluation of nonionic and zwitterionic detergents as membrane protein solubilizers in two-dimensional electrophoresis *Proteomics* **3,** 249-253.
31. Rabilloud, T., Adessi, C., Giraudel, A., and Lunardi, J. (1997) Improvement of the solubilization of proteins in two-dimensional electrophoresis with immobilized pH gradients. *Electrophoresis* **18,** 307-316

6

Oligomeric States of Proteins Determined by Size-Exclusion Chromatography Coupled With Light Scattering, Absorbance, and Refractive Index Detectors

Ewa Folta-Stogniew

Summary

Size-exclusion chromatography (SEC), coupled with "on-line" static laser light scattering (LS), refractive index (RI), and ultraviolet (UV) detection, provides a universal approach for determination of the molar mass and oligomeric state in solution of native proteins as well as glycosylated proteins or membrane proteins solubilized in non-ionic detergents. Such glycosylated proteins or protein–detergent complexes show anomalous behavior on SEC, thus presenting a challenge to determination of molar mass and oligomeric state in solution. In the SEC-UV/LS/RI approach, SEC serves solely as a fractionation step, while the responses from the three detectors are utilized to calculate the molar mass for the polypeptide portion of the native or modified protein. The amount of sugar, lipid, or detergent bound to the polypeptide chain can also be estimated from the SEC-UV/LS/RI analysis.

Key Words: Laser light scattering; size-exclusion chromatography (SEC); molar mass; oligomeric state; glycoproteins; detergent-solubilized membrane proteins.

1. Introduction

Light scattering is a spectroscopic technique for determination of the molar masses of biopolymers in solution *(1)*. Review articles have been published on the theory and application of static light scattering (LS) combined with size-exclusion chromatography (SEC) for determination of the molar masses of proteins in solution *(2–4)*, and for analysis of protein complexes *(4,5)*. This chapter will focus on the most versatile application of light scattering for studying oligomerization of proteins: the SEC-ultraviolet (UV)/LS/refractive index (RI) approach, which can be applied to the analysis of native soluble proteins, proteins with posttranslational modifications (e.g., glycosylated

From: *Methods in Molecular Biology, Vol. 328: New and Emerging Proteomic Techniques*
Edited by: D. Nedelkov and R. W. Nelson © Humana Press Inc., Totowa, NJ

proteins), membrane proteins solubilized with non-ionic detergents, and proteins complexed with polyethylene glycol. Determination of the oligomeric state of such modified proteins or protein–detergent complexes is a challenging task because these proteins behave anomalously during SEC. In contrast, static light scattering combined with refractometry and absorbance offers a more rigorous approach for determination of molar mass and oligomeric state of proteins in solution ***(6)***. Only the sequence of the polypeptide chain needs to be known to determine its oligomeric state using SEC-UV/LS/RI analysis; the degree of modification of the polypeptide chain does not need to be known beforehand. On the contrary, if the chemical nature of the moiety bound to the polypeptide chain is known, the amount of sugar, or polyethylene glycol, or detergent bound to the polypeptide chain can also be estimated from the SEC-UV/LS/RI analysis.

The SEC-UV/LS/RI approach utilizes three detectors: an absorbance detector (UV), a static LS detector, and an RI detector, which are placed in series with an SEC column. The UV detector monitors absorbance at 280 nm, the RI detector monitors changes in refractive index, and the LS detector records the excess of scattered light, whereas SEC serves only as a fractionation step. Because static LS provides only the *weight-average* molar mass of the species in solution, the SEC separation plays an integral role in the overall analysis, although the elution from SEC does not need to correlate with the molar masses of the species being studied. Success requires only that the various oligomeric species present in the sample be physically separated from each other prior to their entrance into the LS detector. The SEC-UV/LS/RI system is calibrated in a buffer of choice by analyzing protein standards. The responses from the three detectors are subsequently used to calculate the molar mass (MM_{pp}) for the polypeptide portion of the modified (or unmodified) protein.

The SEC-UV/LS/RI data are processed utilizing the approach initially described in detail by Hayashi and coworkers ***(6)***, and applied later to the analysis of glycoproteins, proteins modified by polyethylene glycol, and membrane proteins solubilized by non-ionic detergents ***(4,6–10)***.

The molar mass of the polypeptide portion of the native or modified protein, (MM_{pp}), is calculated using the following relationships ***(6)***:

$$MM_{pp} = k_1 \frac{(LS)}{\left(\frac{dn}{dc}\right)_{app} (RI)} \tag{1}$$

$$\left(\frac{dn}{dc}\right)_{app} = k_2 A \frac{(RI)}{(UV)} \tag{2}$$

where:

MM_{pp} is the molar mass of the polypeptide portion of the protein complex;

$\left(\frac{dn}{dc}\right)_{app}$ is the apparent refractive index increment of the protein complex, which relates the changes in the refractive index to changes in the polypeptide concentration, monitored by UV at 280 nm;

LS is the response from the LS detector;

A is the extinction coefficient of the polypeptide, $A^{0.1\%}{}_{280}$, expressed in terms of weight concentration ($mg^{-1}mLcm^{-1}$)—i.e., the absorbance at 280 nm produced by a 0.1% solution (1 mg/mL) of the polypeptide;

RI is the response from the refractive index detector;

UV is the response from the UV detector at 280 nm;

k_1, k_2 are the calibration constants determined from analysis of protein standards during the calibration of the SEC-UV/LS/RI system.

From Eq. 1 and Eq. 2, the molar mass of the polypeptide portion of the protein complex, MM_{pp}, and the oligomeric state, *N*, can be expressed as:

$$MM_{pp} = \frac{K * (LS)(UV)}{A(RI)^2} \tag{3}$$

$$N = \frac{MM_{pp}}{MM} \tag{4}$$

where:

K is the calibration constant for the SEC-UV/LS/RI system as determined from the calibration curve generated during the calibration of the SEC-UV/LS/RI system in the buffer of choice with suitable protein standards;

MM is the molar mass of the polypeptide chain calculated from its sequence.

The factor *A*, i.e., the extinction coefficient at 280 nm expressed in terms of weight concentration, can be computed based on the protein sequence; the molar extinction coefficient, ε_{280} ($M^{-1}cm^{-1}$), is computed as described in Pace et al. ***(11)***:

$$\varepsilon_{280}\ (M^{-1}cm^{-1}) = (\#Trp)(5500) + (\#Tyr)(1490) + (\#Cystine)(125) \tag{5}$$

and ε_{280} ($M^{-1}cm^{-1}$) is converted to $A^{0.1\%}{}_{280}$ ($mg^{-1}mLcm^{-1}$) by dividing ε_{280} ($M^{-1}cm^{-1}$) by the calculated molar mass of the monomeric polypeptide chain, MM (g):

$$A = \varepsilon/MM = [(\#Trp)*(5500) + (\#Tyr)*(1490) + (\#Cystine)*(125)]/MM \quad (6)$$

The approach just described allows determination of the molar mass, MM_{pp}, and the oligomeric state, *N*, of the polypeptide without any prior knowledge of the amount of nonpolypeptide moiety bound to the polypeptide chain. The calibration of the SEC-UV/LS/RI system can be performed easily in a buffer of interest. Systematic errors—e.g., imprecision in detector calibrations or poor monochromator calibration for the UV detector—are accounted for during the calibration step. Because all calculations are carried out without the support of any commercially available software, data analysis is quite laborious, especially if one desires to determine the distribution of MM_{pp} across the eluting peak. For that reason, determination of the MM_{pp} is routinely performed only for the maximum of a peak, using peak heights, or for a selected portion of the eluting peak, using peak volumes. This approach circumvents, however, the need to know the specific refractive index for the protein complex, $(dn/dc)_{complex}$, a parameter that is otherwise needed to determine MM_{pp} using the commercially available software for analysis of light-scattering data, e.g., ASTRA software (Wyatt Corp., Santa Barbara, CA). In principle, $(dn/dc)_{complex}$ can be measured experimentally using a refractometer ***(9)***, but such an approach is usually precluded as a result of limiting amounts of sample. For the majority of conjugated proteins or protein complexes, $(dn/dc)_{complex}$ cannot be calculated beforehand, because information about the extent of modification or the exact composition of the protein complex is not known *a priori*.

The value of $(dn/dc)_{app}$ calculated from Eq. 2 allows estimation of the amount of nonpolypeptide moiety that is bound to the polypeptide chain ***(6)***; namely:

$$\delta = \frac{\left(\frac{dn}{dc}\right)_{app} - \left(\frac{dn}{dc}\right)_{pp}}{\left(\frac{dn}{dc}\right)_{ligand}} \quad (7)$$

where:

δ is the amount (g) of nonpolypeptide components of the protein complex per gram of polypeptide;

$\left(\frac{dn}{dc}\right)_{app}$ is the apparent refractive index increment of the protein complex, $(dn/dc)_{app} = k_2*A*[(RI)/(UV)]$, Eq. 2;

$\left(\frac{dn}{dc}\right)_{pp}$ is the specific refractive index increment for polypeptides in a given buffer (taken from literature values and validated by analysis of protein standards);

$\left(\frac{dn}{dc}\right)_{ligand}$ is the specific refractive index increment for the nonpolypeptide moiety associated with the polypeptide (e.g., sugar in the case of glycoproteins, polyethylene glycol in the case of proteins modified by polyethylene glycol, or lipids and detergents in the case of membrane proteins).

Once the amount of ligand bound to the polypeptide is determined, the total mass of the protein complex can be calculated as:

$$MM_{complex} = MM_{pp}(1 + \delta) \tag{8}$$

Additional data analysis can be performed using commercially available ASTRA software (Wyatt Corp., Santa Barbara, CA), which determines molar mass (MM) by solving the equation that relates the excess scattered light, measured at several angles, to the concentration of solute and the weight-average molar mass *(1)*. Please refer to **http://info.med.yale.edu/wmkeck/biophysics/Astra2a.htm**. Calculation of MW by ASTRA for details on ASTRA calculations. The specific refractive index increment for the entire protein complex needs to be known to utilize ASTRA, and can be calculated at this point as a weight average of the specific refractive index increments of the parent components ***(12)***:

$$\left(\frac{dn}{dc}\right)_{complex} = \frac{MM_{pp}}{MM_{complex}}\left(\frac{dn}{dc}\right)_{pp} + \frac{(MM_{complex} - MM_{pp})}{MM_{complex}}\left(\frac{dn}{dc}\right)_{ligand} \tag{9}$$

When the literature values for either $(dn/dc)_{pp}$ or $(dn/dc)_{ligand}$ or both these constants are used, the computed value for the $(dn/dc)_{complex}$ is obviously an approximation that introduces additional uncertainty into the MM determination. In ASTRA calculations, the concentration is estimated from the RI signal, and the dn/dc value contributes inversely to the computation of MM; thus, the error in determination of MM due to uncertainly in dn/dc will be as large as the error in the dn/dc estimation. For example, the dn/dc value for the polypeptide chain varies depending on buffer composition, from 0.18 to 0.20 mL/g, giving an average value of 0.19 ± 5%, *(8)*; such a ±5% error in the dn/dc value will propagate as a corresponding ±5% error in MM determination. Nonetheless, there are strong advantages in using ASTRA for additional data analysis: ASTRA utilizes scattering signals collected at multiple angles, which increases the precision of the MM determination; secondly, ASTRA computes MM across the entire eluting peak (typically 100–200 measurements within one peak) and thus detects whether the protein is monodisperse, i.e., homogeneous with respect to molar mass.

The examples shown in **Figs. 1** and **2** result from the SEC-UV/LS/RI analysis of two well characterized membrane proteins with known crystal structures: a heptameric hemolysin (monomeric MM of 33 kDa) from *Staphylococcus aureus*,

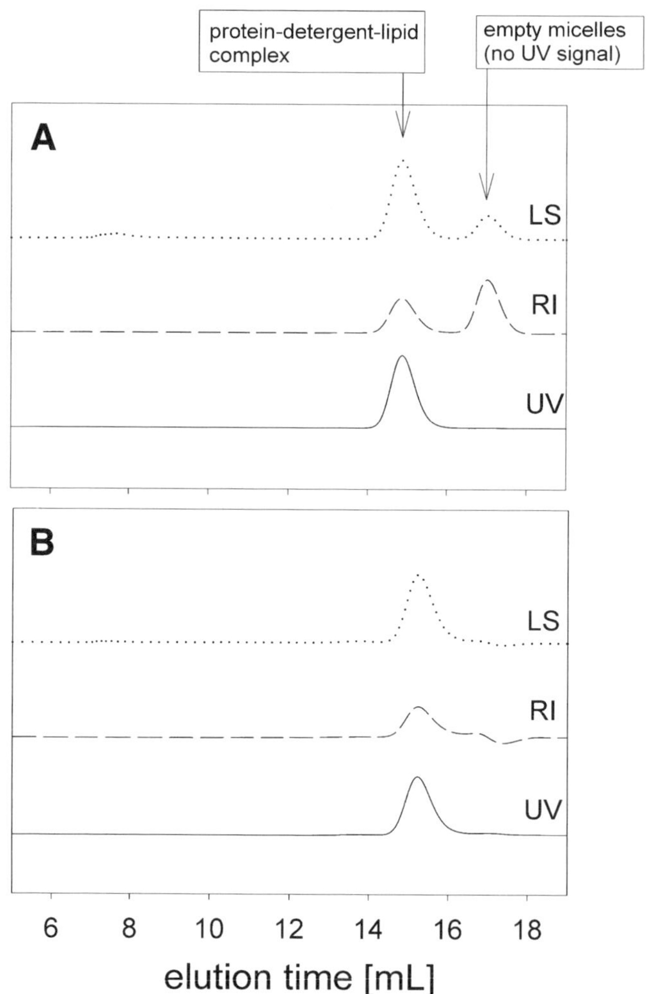

Fig. 1. Elution profiles of the LamB maltoporin and α-HL hemolysin monitored by laser light scattering (LS) (the trace shown was recorded at 90-degree angle), absorbance at 280 nm (ultraviolet [UV]), and refractive index (RI). A Superose 6 HR 10/30 column was equilibrated in 20 m*M* HEPES (pH 7.4), 200 m*M* NaCl, 1 m*M* ethylenediamine tetraacetic acid, 5 m*M* L-glutamic acid, 2 m*M* (0.05%) dodecyl maltoside (C12M), prior to injecting approx 50 µg of purified hemolysin, α-HL, and approx 100 µg of purified maltoporin, LamB. Traces collected by the LS (dotted line), RI (dashed line), and UV (solid line) detector are shown for maltoporin, LamB (**A**), and for hemolysin, α-HL (**B**).

α-HL, and a trimeric maltoporin (monomeric MM of 47 kDa) from *E. coli*, LamB, ***(13,14)***. These proteins are soluble and stable in aqueous buffer supplemented with non-ionic detergent, in which they exist as complexes of detergent and lipids associated with the polypeptide chain ***(10)***. The SEC-UV/LS/RI analysis allows deter-

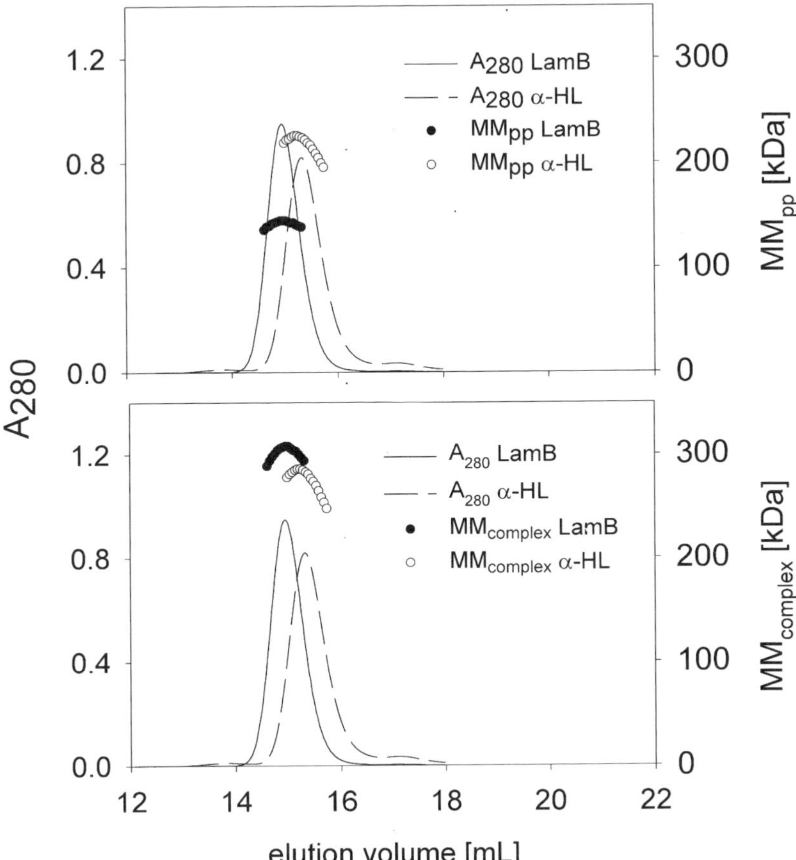

Fig. 2. Molar mass distribution plots from size-exclusion chromatography (SEC)-ultraviolet (UV)/light scattering (LS)/refractive index (RI) data for maltoporin, LamB, and hemolysin, α-HL. Molar masses for the polypeptide portion of the protein complexes (MM_{pp}) are plotted in **A**; molar masses for the protein complexes ($MM_{complex}$) are plotted in **B**. Lines correspond to UV traces of the proteins eluting from the SEC column, monitored at 280 nm; maltoporin, LamB, UV trace—solid line; molar mass—filled circles; hemolysin, α-HL, UV trace—dashed line; molar mass open circles (for clarity, only every tenth measurement of molar mass across the eluting peak is plotted).

mination of the molar mass and oligomeric state in solution and provides estimates of the amount of detergents and lipids bound to the polypeptide chain. Because these two proteins vary greatly in their hydrophobic surfaces, they are expected to bind different amounts of detergent. Additionally, this example illustrates clearly that SEC alone is not a reliable method to determine the oligomeric state of modified proteins, and detergent-solubilized membrane proteins in particular, because their elution from SEC is dependent on the total mass of the polypeptide–detergent–lipids complex rather than on their oligomeric state and the molar mass of the polypeptide itself ***(6,7,9,10)***.

The following protocol outlines (1) setup and equilibration of the SEC-UV/LS/RI system; (2) calibration of the SEC-UV/LS/RI system in the buffer of choice; (3) sample preparation, data collection, and processing; (4) determination of the molar mass of the polypeptide and its oligomeric state; and (5) determination of the amount of the nonpolypeptide moiety bound to the polypeptide; which leads to (6) determination of the molar mass for the entire protein complex/conjugated protein. For native proteins, which are known *a priori* to be pure polypeptide chains, **Subheadings 3.5.** and **3.6.** are omitted during data processing, and $(dn/dc)_{pp}$ is substituted for $(dn/dc)_{complex}$ during the ASTRA analysis under **Subheading 3.7.**

2. Materials

2.1. SEC/LS/UV/RI System

2.1.1. Instruments (see **Note 1**)

1. High-performance liquid chromatography (HPLC) pump capable of delivering a flow rate of 0.3 to 1.0 mL/min with low pulsation (Alliance 2695, Waters Corporation, Milford, MA).
2. In-line solvent filter with 0.1-µm pore size, installed between the pumps and SEC column (*see* **Note 2**).
3. In-line sample filter with 2-µm pore size, installed between the injector and SEC column (*see* **Note 2**).
4. UV detector for in-line use (996 PDA, Waters Corporation, Milford, MA).
5. RI detector for in-line use (OptiLab DSP, Wyatt Technology, Santa Barbara, CA).
6. Static LS detector for in-line use (DAWN EOS, Wyatt Technology, Santa Barbara, CA).
7. Computer and software for data collection; preferably software capable of collecting data from all three detectors simultaneously (ASTRA software, Wyatt Technology, Santa Barbara, CA; Millennium software, Waters Corporation, Milford, MA).

2.1.2. SEC Column

High-performance size-exclusion chromatography column appropriate for fractionating the samples analyzed (Superose 6, HR 10/30, GE Healthcare-Amersham Biosciences, Piscataway, NJ; use **Table 1** guidelines regarding column selection; *see* **Note 3**).

2.2. Reagents and Supplies

1. Filters with low protein-binding properties and pore size of 0.22 µm (*see* **Note 4**).
2. Protein standards: five proteins with MM_{pp} spanning the range of the expected MM_{pp} of the sample to be analyzed (*see* **Note 5**).

Table 1
Sample and Column Requirements for SEC-UV/LS/RI Analysis

	Optimal amount of protein µg [10^{-6} g]			
Column[a]	MM >200 kDa	MM 40–200 kDa	MM 10–40 kDa	MM <10 kDa
Superose 6 HR 10/30	**50**	**50–100**	Not suitable	Not suitable
Superdex 200 HR 10/30	50	**50–100**	**100–200**	Not suitable
Superdex 75 HR 10/30	Not suitable	50–100	**100–200**	Not suitable
Superdex peptide HR 10/30	Not suitable	Not suitable	Not suitable	**400–800**

[a] the Superose/Superdex columns from the HR 10/30 series were extensively tested by the author in a variety of buffer conditions (including buffers supplemented with various detergents or denaturants) as suitable for the SEC-UV/LS/RI approach; shown in bold type are the optimal column matches for a given MM range (as expected for a given oligomeric state). SEC, size-exclusion chromatography; UV, ultraviolet; LS, light scattering; RI, refractive index; MM, molar mass.

3. Buffer: aqueous buffer compatible with the SEC column requirements; for the majority of SEC media, 150 m*M* salts must be present to prevent electrostatic interactions with the column's matrix (for the analysis of α-HL and LamB: 20 m*M* HEPES, 200 m*M* NaCl, 1 m*M* ethylenediamine tetraacetic acid [EDTA]; pH 7.4, 5 m*M* L-glutamic acid, 2 m*M* (0.05%) dodecyl maltoside, C12M). The buffer should be filtered through a 0.1-µm filter.
4. Sample: 50–500 µg of protein sample (*see* **Table 1** for guidelines regarding optimal sample amounts) in a volume that corresponds to less than 3% of the total volume of the SEC column (with 1% being the preferred loading volume).

3. Methods

3.1. System Setup and Calibration of Detectors

The detectors should be connected in the following order: UV, LS, and RI. The RI detector should be last in this series, because its cell is fragile and it cannot withstand high pressures. Tubing with an inner diameter of 0.25 mm should be used throughout the system to minimize the delay volumes between detectors and minimize band-broadening effects. The LS detector should be calibrated according to the protocol provided by the manufacturer; the calibration remains valid for a couple of years unless changes are made in the laser or photodiodes (provided that the flow cell is kept clean). The RI detector should be calibrated to convert changes in voltage (Volts) to changes in refractive index (dn); this calibration remains valid for 1 yr (*see* **Note 6**).

3.2. System Equilibration

1. Equilibrate the SEC-UV/LS/RI system in the buffer of choice; turn on the RI detector and pump the buffer through the RI detector in "purge" mode for at least 12 h.
2. Turn on the LS and UV detectors; the UV detector requires 30 min to warm up the lamp, and the LS detector requires 30 min to warm up its electronics.
3. Stop the purging flow on the RI detector, zero the RI detector, and start monitoring baselines.
4. Check the baseline quality and stability; the noise level in the LS baseline monitored at 90 degrees should not exceed 4 mV (with the goal being to keep the noise at less than 2 mV) at the highest sensitivity setting of the LS detector; the drift in the RI baseline should not exceed 0.5% of full scale per 20 min (*see* **Note 7**).

3.3. Calibration of SEC/LS/UV/RI System in the Buffer of Interest

1. Dissolve 500 μg of protein standard in 400 μL of running buffer (use at least five different, individually prepared protein standards; *see* **Note 5**).
2. Filter the solutions of protein standards through a 0.22-μm low protein-binding filter.
3. Individually inject 200 μg of each protein standard and collect UV, LS, and RI data.
4. Baseline-correct the signals from all three detectors and calculate volumes or heights of the UV, LS, and RI peaks for each data set.
5. Calculate theoretical extinction coefficient, A, for each protein analyzed according to Eq. 6.
6. Create a calibration curve by plotting MM_{pp} of the standards (according to their known oligomeric states) as a function of $(LS)*(UV)/[A*(RI^2)]$.
7. Calculate the calibration constant, K, from least-square linear regression of Eq. 3.

3.4. Determination of the Oligomeric State of the Sample Protein

1. Filter the protein sample through a 0.22-μm low protein-binding filter.
2. Inject an appropriate amount of sample (*see* **Table 1** for optimal amounts of sample) and record signals from all three detectors (an example of three-detector monitoring is shown in **Fig. 1**).
3. Compute the extinction coefficient, A, for the protein analyzed using Eq. 6.
4. Baseline-correct the signal from all three detectors.
5. Measure volume (or height) of the UV, LS, and RI peaks.
6. Calculate molar mass of the polypeptide portion of the protein or protein complex, $MM_{pp} = [K*(LS)*(UV)]/[A*(RI^2)]$ (Eq. 3), using the calibration constant, K, computed in **Subheading 3.3.7.** (results for α-HL and LamB are reported in **Table 2** as MM_{pp}).
7. Determine the oligomeric state from the ratio of MM_{pp} and MM of the polypeptide chain as calculated from the sequence (Eq. 4; results for α-HL and LamB are reported in **Table 2**).

Table 2
Results of SEC-UV/LS/RI Analysis of a Heptameric Hemolysin from *Staphylococcus aureus*, α-HL, and a Trimeric Maltoporin From *Escherichia coli*, LamB

Protein MM of monomer (kDa)	Molar mass for polypeptide MM_{pp} (kDa): Determined from SEC-UV/LS/RI	Molar mass for polypeptide MM_{pp} (kDa): Expected	Oligomeric state: Determined from SEC-UV/LS/RI	Oligomeric state: Expected	δ (g/g) Amount of detergent/lipids per gram of polypeptide	$MM_{complex}$ (kDa)
α-HL MM = 33 kDa	215 ± 20[a,b]	231	6.5 ± 0.6[b]	7	0.26	271
LamB MM = 47 kDa	141 ± 3	141	3.00 ± 0.06	3	1.2	310

[a]Average from three or four measurements for LamB and α-HL, respectively; errors represent one standard deviation.

[b]For SEC-UV/LS/RI analysis of modified proteins, typical error is approx ±10%, depending mostly on the amount of sample analyzed, stability of SEC-UV/LS/RI system, and precision of estimation of $(dn/dc)_{ligand}$; thus, for heptamers and higher-order oligomers, the uncertainty in the determination of MM_{pp} may reach the mass of the monomer.

SEC, size-exclusion chromatography; UV, ultraviolet; LS, light scattering; RI, refractive index; MM, molar mass.

3.5. Determination of the Amount of a Nonpolypeptide Moiety in a Protein Complex

1. Calculate k_2 for each protein standard analyzed using Eq. 2 (use 0.187 mL/g for $[dn/dc]_{app}$ for protein standards; *see* **Note 8**); compute average value of k_2.
2. Compute $(dn/dc)_{app}$ for the protein sample according to Eq. 2, using the average k_2 value.
3. Calculate δ according to Eq. 7, using 0.187 mL/g for $(dn/dc)_{pp}$ and an appropriate $(dn/dc)_{ligand}$ (*see* **Note 8**; results of this analysis for α-HL and LamB are reported in **Table 2**).

3.6. Determination of the Molar Mass of the Protein Complex or Conjugated Protein

1. Calculate $MM_{complex}$ as $MM_{pp}*(1+\Delta)$ (Eq. 8; results for α-HL and LamB are reported in **Table 2**).
2. Compute $(dn/dc)_{complex}$ according to Eq. 9.

3.7. Determination of Monodispersity

1. Use the $(dn/dc)_{complex}$ during processing of the SEC-UV/LS/RI data using ASTRA software.
2. Generate a molar mass distribution plot (an example of such a plot for α-HL and LamB is shown in **Fig. 2**).
3. Check the distribution of $MM_{complex}$ across the eluting peak; for a monodisperse sample, it should vary by no more than ±5% for the middle portion of the peak that is above half-height (thus, it is not significantly affected by band-broadening effects).

4. Notes

1. The data for the examples shown were collected using an SEC-UV/LS/RI system consisting of an HPLC system, Alliance 2960, (Waters Corp., Milford, MA) equipped with autosampler, photodiode array (PDA) UV/VIS detector (996 PDA, Waters Corp., Milford, MA); differential refractometer (OPTI-Lab, Wyatt Corp., Santa Barbara, CA); and static, multiangle laser LS detector (DAWN-EOS, Wyatt Corp., Santa Barbara, CA). Two software packages were used during its operation; Millennium software (Waters Corp., Milford, MA) controlled the HPLC operation and data collection from the multi-wavelength UV/VIS detector, whereas ASTRA software (Wyatt Corp., Santa Barbara, CA) collected data from the RI detector and the LS detector, and also recorded the UV trace at 280 nm sent from the PDA detector. However, other UV, RI, and LS detectors can be utilized, and the data can be processed even in the absence of commercially available software for LS data processing (a single-angle LS detector is suitable for analysis of proteins with molar masses up to approx 500 kDa). Data analysis is dependent only on the estimation of volumes (or heights) of peaks generated in various detectors.
2. The 0.1-μm in-line filter placed between pumps and the injector retains any particles that are shed from the pump's head. An additional in-line filter (2-μm PEEK

frit) with small dead volume is installed between the injector and the SEC column; this filter traps protein aggregates that are formed during the injection step and that may result from exposing the protein sample to high pressure; this filter substantially increases the lifetime of the SEC column. These filters are replaced whenever the system's operating pressure increases by more than 5%.

3. The SEC columns are rated for specific ranges of MMs that can be fractionated on a given column; for example, the Superdex columns (Amersham Pharmacia) cover three ranges: Superdex peptide is suitable for fractionation of globular proteins with MMs in the range of 100 Da to 7 kDa; Superdex 75, for MMs ranging from 3 to 70 kDa; and Superdex 200 for MMs of 10 to 600 kDa. During SEC-UV/LS/RI analysis, it is preferable for the protein to elute at the lower end of the fractionation range and to be well separated from the particles or high-MM aggregates that elute at the initial portion of the chromatogram. Thus, for a polypeptide of 50 kDa, Superdex 200 is a better choice than Superdex 75 (*see* **Table 1**). Because the aggregates usually represent less than 0.1% of the total eluting mass, they produce a very weak signal in the mass detectors (UV or RI). Nonetheless, as a result of their high MM, the aggregates or particles generate a strong LS peak which tails into the initial portion of the chromatogram (even when the mass eluting in the void volume is just a fraction of 1% of the total sample mass).

   There are many SEC columns suitable for SEC-UV/LS/RI systems. The Superdex and Superose HR 10/30 series columns are routinely used by our laboratory. Other commonly used columns are the TosoHaas TSK-GEL series SEC columns. The silica-based TSK-GEL columns provide excellent resolution, but are limited in chemical stability to the pH range from 2.0 to 8.0. Although they provide lower resolution than silica-based columns, the polymer-based TSK-GEL columns are a very good choice for samples with high MMs. Laboratory-packed columns can also be used when a specific medium is required to minimize the interactions with the SEC matrix. The critical factor in a column's compatibility with the SEC-UV/LS/RI analysis is the quality of the LS baseline after a new column has been conditioned. Every new column sheds particles, and the noise in the LS baseline is initially high, but after 24–72 h of continuous pumping, the baseline noise should decrease to less than 3 mV when recorded for the 90-degree detector at the highest sensitivity setting. A major challenge for SEC column selection is analysis of large aggregates with MM above $1 \cdot 10^6$Da (1 MDa), for which field-flow fractionation (reviewed in **ref. *15***), would be a preferred method of fractionation.

4. For protein samples in aqueous buffers, the 0.1- or 0.22-μm Durapore®, low protein-binding membrane (Millipore, Bedford, MA) has proven to be an excellent choice; centrifugal filter units have low dead volume and are thus preferred over syringe filters. For samples of detergent-solubilized membrane proteins, wetting the filter surface with the detergent-containing buffer may prevent nonspecific binding and improve recovery.

5. Protein standards that are routinely used for SEC calibration are good choices for SEC/LS analysis (the individually sold standards, not the premixed sets). The list of 16 commercially available proteins, with MM_{pp} ranging from 6.5 kDa to 475 kDa, that our laboratory analyzed by SEC-UV/LS/RI system is posted at:

http://info.med.yale.edu/wmkeck/biophysics/Sumtaba.htm. Please note that albumins (like bovine serum albumin, or ovalbumin), which are routinely used as protein standards, are lipid-carrying proteins and can bind some of the detergents used for solubilization of membrane proteins.

6. Calibration of the RI detector provides conversion of the measured voltage changes, dV, to changes in refractive index, dn. The response, in volts, of the RI detector is recorded for six to eight aqueous solutions of NaCl of known concentrations (a typical range would be 0.1 to 1 mg/mL). At a dn/dc value of 0.174 mL/g for NaCl (633 nm), the changes in concentration are converted to changes in refractive index, dn, and plotted as a function of changes in voltage; the calibration constant is then obtained from the slope.
7. Proper equilibration of the SEC-UV/LS/RI system is critical for obtaining high precision in the MM determination; the baselines must be monitored for at least 60 min to ensure that there is no substantial drift in RI signal, and that the LS baseline is stable and free of noise. Because the response of the LS detector determines the lower limit of detection, low noise is critical when working with a low amount of sample (e.g., 20 µg of a 40-kDa macromolecule). When not in use, the SEC-LS/UV/RI system should be stored with 20% ethanol continuously recirculating through it. For long-term storage, the LS cell should be filled with methanol filtered through a 0.02-µm filter.
8. Please note that the error in the $(dn/dc)_{complex}$ results mostly from the uncertainty in $(dn/dc)_{ligand}$, because $(dn/dc)_{pp}$ can be estimated with high precision during ASTRA analysis of SEC-UV/LS/RI data collected for standards; starting with the commonly used value of 0.187 mL/g ***(2,4,6,8,10)***, the iterative process leads to a single $(dn/dc)_{pp}$ value within the range of 0.18 to 0.20 mL/g, providing thc best match between the expected and measured MM_{pp} for all standards analyzed and taking into account their expected oligomeric state (*see* comment in **Note 5** for albumins). Alternatively, the $(dn/dc)_{pp}$ can be measured using any standard protein that has been thoroughly equilibrated with the buffer of interest (i.e., by running a preparative SEC or carrying out extensive dialysis). For the specific refractive index of polypeptide, $(dn/dc)_{pp}$, a value of 0.187 mL/g is widely used ***(2–4,6,8,10)***; for sugars, $(dn/dc)_{sugar}$ of 0.140 mL/g has been determined ***(8,16)***; for polyethylyne glycol, $(dn/dc)_{PEG}$ can be measured using an RI detector ***(3,4,8)***; for lipids and detergents, a $(dn/dc)_{lipids/detergent}$ of 0.134 mL/g is commonly used ***(6,8–10)***.

Acknowledgments

This research has been funded in part with federal funds from National Heart, Lung and Blood Institute/National Institutes of Health (NIH) contract N01-HV-28186 and National Institute on Drug Abuse/NIH grant 1 P30 DA018343-01. I thank Olga Boudker and Dinesh Yernool (Dr. Eric Gouaux's laboratory, Columbia University, New York, NY) for providing the membrane proteins and for fruitful discussions regarding the behavior of detergent-solubilized membrane proteins. I gratefully acknowledge Dr. Ken Williams (W. M.

Keck Biotechnology Resource Laboratory, Yale University, New Haven, CT) for critically reviewing the manuscript.

References

1. Wyatt, P. J. (1993) Light scattering and the absolute characterization of macromolecules. *Analytica Chimica Acta* **272,** 1–40.
2. Folta-Stogniew, E. and Williams, K. R. (1999) Determination of molecular masses of proteins in solution: implementation of an HPLC size exclusion chromatography and laser light scattering service in a core laboratory. *J. Biomol. Tech.* **10,** 51–63.
3. Takagi, T. (1990) Application of low-angle laser light scattering detection in the field of biochemistry: review of recent progress. *J. Chromatog. A* **506,** 409–416.
4. Wen, J., Arakawa, T., and Philo, J. S. (1996) Size-exclusion chromatography with on-line light-scattering, absorbance, and refractive index detectors for studying proteins and their interactions. *Anal. Biochem.* **240,** 155–166.
5. Mogridge, J. (2004) Using light scattering to determine the stoichiometry of protein complexes. *Meth. Mol. Biol.* **261,** 113–118.
6. Hayashi, Y., Matsui, H., and Takagi, T. (1989) Membrane protein molecular weight determined by low-angle laser light-scattering photometry coupled with high-performance gel chromatography. *Meth. Enzymol.* **172,** 514–528.
7. Hayashi, Y., Takagi, T., Maezawa, S., and Matsui, H. (1983) Molecular weights of [alpha][beta]-protomeric and oligomeric units of soluble (Na+, K+)-ATPase determined by low-angle laser light scattering after high-performance gel chromatography. *Biochim. Biophys. Acta* **748,** 153–167.
8. Kendrick, B. S., Kerwin, B. A., Chang, B. S., and Philo, J. S. (2001) Online size-exclusion high-performance liquid chromatography light scattering and differential refractometry methods to determine degree of polymer conjugation to proteins and protein-protein or protein-ligand association states. *Anal. Biochem.* **299,** 136–146.
9. Wei, Y., Li, H., and Fu, D. (2004) Oligomeric state of the *Escherichia coli* metal transporter YiiP. *J. Biol. Chem.* **279,** 39,251–39,259.
10. Yernool, D., Boudker, O., Folta-Stogniew, E., and Gouaux, E. (2003) Trimeric subunit stoichiometry of the glutamate transporters from Bacillus caldotenax and Bacillus stearothermophilus. *Biochemistry* **42,** 12,981–12,988.
11. Pace, C. N., Vajdos, F., Fee, L., Grimsley, G., and Gray, T. (1995) How to measure and predict the molar absorption coefficient of a protein. *Protein Sci.* **4,** 2411–2423.
12. Prochazka, O. and KratochvÌl, P. (1980) Light scattering in multicomponent solutions a general equation. *J. Polymer Sci.* **18,** 2369–2377.
13. Song, L., Hobaugh, M. R., Shustak, C., Cheley, S., Bayley, H., and Gouaux, J. E. (1996) Structure of staphylococcal alpha-hemolysin, a heptameric transmembrane pore. *Science* **274,** 1859–1865.
14. Schirmer, T., Keller, T. A., Wang, Y. F., and Rosenbusch, J. P. (1995) Structural basis for sugar translocation through maltoporin channels at 3.1 A resolution. *Science* **267,** 512–514.
15. Giddings, J. C. (1993) Field-flow fractionation: analysis of macromolecular, colloidal, and particulate materials. *Science* **260,** 1456–1465.

16. D'ambra, A. J., Baugher, J. E., Concannon, P. E., Pon, R. A., and Michon, F. (1997) Direct and indirect methods for molar-mass analysis of fragments of the capsular polysaccharide of *Haemophilus influenzae* type b. *Anal. Biochem.* **250,** 228–236.

7

Surface Plasmon Resonance Imaging Measurements of Protein Interactions With Biopolymer Microarrays

Terry T. Goodrich, Alastair W. Wark, Robert M. Corn, and Hye Jin Lee

Summary

The surface-sensitive optical technique of surface plasmon resonance (SPR) imaging is an ideal method for the study of affinity binding interactions of unlabeled biological molecules in a multiplexed format. This approach has been widely applied to monitor DNA–DNA, DNA–RNA, peptide–protein, and protein–protein interactions as well as surface enzyme reactions. The success of SPR imaging measurements relies on the robust attachment of biomolecules in an array format. In this chapter, we introduce two different surface attachment chemistries that covalently immobilize DNA and peptides onto gold surfaces through the modification of self-assembled alkanethiol monolayers. Array fabrication approaches for the creation of individually addressable elements through the use of either gold dot patterns or polydimethylsiloxane (PDMS) microchannels are detailed. The utility of SPR imaging for the study of protein interactions is demonstrated with two biological systems: the binding of response regulator proteins, VanR and OmpR, onto a DNA array, and the interaction of S protein with an array of S-peptide variants. Furthermore, the application of real-time SPR imaging to the multiplexed determination of S-protein adsorption/desorption kinetics is described.

Key Words: Surface plasmon resonance (SPR) imaging; DNA microarrays; peptide microarrays; polydimethylsiloxane (PDMS) microfluidic channels; VanR and OmpR; S protein; S peptide.

1. Introduction

The ability to detect proteins in an array format would be highly beneficial for many applications, such as epitope mapping *(1)*, protein identification *(2,3)*, screening of potential protein-binding inhibitors, and protein-binding kinetics *(4,5)*. Typically, protein-binding studies make use of fluorescent tags or radiolabels for detection purposes. However, labeling presents the potential problem of altering the protein's active binding site and thus its biological activity. The technique of surface plasmon resonance (SPR) is ideally suited to protein-binding

From: *Methods in Molecular Biology, Vol. 328: New and Emerging Proteomic Techniques*
Edited by: D. Nedelkov and R. W. Nelson © Humana Press Inc., Totowa, NJ

studies, as affinity interactions are detected by local changes in the index of refraction, and thus tagging of the target molecule is not required.

Surface plasmons are surface electromagnetic waves that propagate parallel to a metal/dielectric interface and are generated when energy from a photon of p-polarized light incident onto a thin metal film, such as gold (Au), is coupled into oscillating modes of free electron density. Surface plasmons have a maximum intensity at the metal/dielectric interface and decay exponentially away from the interface. This gives SPR a sensing depth of only 200 nm, making it a surface-sensitive technique ***(6,7)***.

SPR measurements can be made in a variety of formats including scanning-angle SPR, Fourier transform (FT) SPR, and SPR imaging. In a scanning-angle SPR experiment, the change in percent reflectivity is measured as a function of incident angle, while the wavelength of light is held fixed. This is the most common SPR method and has been popularized by the commercially available BIAcore instrument. In FT-SPR, the wavelength of light is scanned as the angle of incidence is held fixed. Both of these techniques are single-channel measurements and cannot be used in a high-throughput format. SPR imaging is a fixed-angle and wavelength technique where arrays of molecules can be analyzed in parallel over the entire array surface.

In an SPR imaging experiment, the intensity of reflected light from the metal surface is collected at an optimum angle just off of the plasmon angle. When a molecule is adsorbed onto the surface, the SPR curve will shift to a higher angle, as a result of a change in the index of refraction at the surface. This is observed as a change in the light intensity reflected from the metal surface, and is captured by a charge-coupled device (CCD) camera. Changes in percent reflectivity in individual array elements are determined by subtracting an image taken before (reference) from an image taken after exposure to analyte. In equilibrium measurements, sufficient time must be allowed between the two images to ensure that the reaction is complete. Alternatively, the surface reaction can be continually monitored in real time to provide multiplexed kinetic information on processes such as adsorption and desorption.

The key to applying SPR imaging for the study of biomolecular interactions is the development of robust surface-attachment chemistries. In addition, array fabrication procedures are necessary to create surfaces with spatially resolved detection elements for high-throughput studies. In this chapter, two covalent attachment strategies are presented for the creation of both short single-stranded oligonucleotide and peptide microarrays on thin gold films. SPR imaging is then employed in conjunction with DNA arrays to evaluate the binding specificity of the response regulator proteins VanR and OmpR to various DNA promoter regions ***(8)***. Finally, both equilibrium and real-time measurements of differential S-protein binding to an array of S-peptide variants ***(9)*** are demonstrated.

2. Materials

2.1. Thiol-Modified DNA Purification and Preparation

1. 1 *M* triethylammonium acetate (TEAA) (pH 7.01) is prepared by placing 800 mL of water in a 1-L flask (*see* **Note 1**). Place the flask in a fume hood, and add 140 mL of triethylamine while stirring on ice. Slowly add acetic acid over several hours while stirring on ice to adjust the pH to 7.01. Store at 4°C.
2. Buffer A: 4% acetonitrile (high-performance liquid chromatography (HPLC)-grade), 10% TEAA, 86% water.
3. Buffer B: 50% acetonitrile (HPLC-grade), 10% TEAA, 40% water.
4. 50 m*M* phosphate buffer (pH 8.4). Store at room temperature.
5. 200 m*M* dithiothreitol (DTT) in 50 m*M* phosphate buffer (pH 8.4).
6. 100 m*M* triethanolamine hydrochloride (TEA) buffer at both pH 7.0 and pH 8.0. Store at room temperature.
7. 4 mg of Ellman's reagent (0.4% 5,5′,-dithiobis[2-nitrobenzoic acid]) dissolved in 1 mL of TEA buffer (pH 8.0) and used immediately. This solution is light sensitive and should be made immediately before use.
8. C6 S-S 5′ thiol-modified DNA used for array fabrication is obtained commercially (*see* **Note 2**).

2.2. Thiol-Modified DNA Array Fabrication

1. 100 m*M* triethanolamine hydrochloride (TEA) buffer at both pH 7.0 and pH 8.0. Store at room temperature.
2. A 9% Cytop (Bellex International, Wilmington, DE) solution is diluted to a working concentration of 1.5% using CT-180 perfluorosolvent (Bellex International, Wilmington, DE).
3. A 1 m*M* solution of 11-mercaptoundeclyamine (MUAM, Dojindo Laboratories) is prepared by dissolving 1.7 mg in approx 7 mL of absolute ethanol. Sulfosuccinimidyl 4-(*N*-malemidomethyl) cyclohexane-1-carboxylate (SSMCC, Pierce) is dissolved at a 1 m*M* concentration in 100 m*M* TEA buffer (pH 7.0).
4. 1 m*M* solutions of thiol-modified DNA in 100 m*M* TEA buffer (pH 7.0).
5. A thin stainless steel mask containing 700-µm-diameter holes spaced 550 µm apart (edge to edge) is used to vapor deposit gold spots onto Cytop-coated SF-10 (Schott Glass) glass slides.

2.3. PDMS Microfluidic Channels

1. Silicon wafers (3-in diameter; International Wafer Services Inc.) are used for the fabrication of three-dimensional (3D) silicon wafer masters for the production of PDMS microfluidic channels.
2. Rigid chrome masks containing the desired microfluidic features are designed using a CAD program and created using e-beam photolithography. High-resolution transparencies containing the desired microfluidic features can be used as an alternative to the chrome masks.
3. Negative photoresist (SU-8 50, Microlithography Chemical Corp.).
4. Propylene glycol methyl ether acetate (Developer, Microlithography Chemical Corp.).

5. Tridecafluoro-1,1,2,2-tetrahydroctyl-1-trichlorosilane (Gelest, Inc.).
6. PDMS prepolymer and curing agent (Sylgard 184, Dow Corning).

2.4. Peptide Array Fabrication

1. A 1 m*M* solution of MUAM is prepared by dissolving 1.7 mg in approx 7 mL of absolute ethanol.
2. A 6.4 m*M* solution of *N*-succinimidyl 3-(2-pyridyldithio) propionamido (SPDP; Pierce) is made in a 1:1 ratio of DMF and 0.1 *M* phosphate-buffered saline (PBS) solution (pH 7.4) immediately prior to use.
3. 2 m*M* solutions of cysteine-modified peptides are prepared in PBS buffer (pH 7.4) prior to immobilization.
4. 2 mg of *N*-hydroxysuccinimidyl (NHS) ester of methoxypoly (ethylene glycol) propionic acid MW 2000 (PEG-NHS, Nektar) is dissolved in 250 µL of TEA buffer (pH 8.0). Prepare fresh immediately before use.

2.5. Affinity Binding Measurements

1. Response regulator protein binding buffer: 10 m*M* phosphate buffer (pH 7.1), 5 m*M* $MgCl_2$, and 100 m*M* NaCl.
2. Bacterial response regulator proteins, VanR and OmpR, are phosphorylated in 100 µL of 50 m*M* HEPES (pH 7.2), 5 m*M* $MgCl_2$, and 50 m*M* acetyl phosphate prior to use.
3. 8 *M* urea, stored at room temperature.
4. S-protein binding buffer: 10 m*M* phosphate buffer (pH 7.4), 2.7 m*M* KCl, and 137 m*M* NaCl.

3. Methods

Successful SPR imaging measurements of protein interactions with biopolymer microarrays require three key components: robust surface chemistry for tethering biopolymers onto a gold surface, array fabrication methods, and surface biochemistry. A fabrication methodology that utilizes alkanethiol chemistry and gold patterned glass substrates to create arrays of DNA molecules is presented first. These arrays are used in conjunction with SPR imaging to measure the relative binding affinity of the bacterial response regulator proteins, VanR and OmpR *(8)*, to various DNA sequences. In addition, an alternative approach employing the use of PDMS microfluidic networks *(10)* to prepare arrays of S-peptide variants onto chemically modified gold surfaces is discussed *(9)*. SPR imaging is used for the acquisition of both equilibrium and real-time measurements of S-protein binding to the prepared arrays.

3.1. SPR Imaging of DNA–Protein Interactions

3.1.1. DNA Purification and Quantification

1. Dilute the thiol DNA with 50 µL of 50 m*M* phosphate buffer (pH 8.4) for every 10 OD_{260} of DNA concentration.

2. Take a 50-μL aliquot of the diluted DNA and mix it with 50 μL of a freshly prepared 200 m*M* solution of DTT diluted in phosphate buffer (pH 8.4). Let it react for 30 min to cleave the disulfide bond of the C6 thiol S-S modifier.
3. Purify the DNA using reverse-phase binary HPLC (*see* **Note 3**).
4. Collect the purified DNA in an Eppendorf tube. Elution time for the DNA occurs in approx 20–25 min.
5. Dry the DNA using a Speed-vac until all of the solution has evaporated.
6. Resuspend the dried DNA in TEA buffer (pH 7.0). Use approx 5 μL of buffer to resuspend the DNA collected from each HPLC purification run.
7. To calculate the concentration of the resuspended DNA, set the detection wavelengths of an ultraviolet (UV)-Vis spectrometer to 260 and 412 nm, and blank the system by placing 59.6 μL of TEA buffer (pH 8.0) into a cuvette.
8. Add 0.4 μL of Ellman's reagent to the cuvette and take a measurement. This will be used as the 100-fold dilution reference.
9. Add an additional 540 μL of TEA buffer (pH 8.0) to the cuvet and take another measurement. This will be used as the 1000-fold dilution reference.
10. In a clean cuvet, add 59 μL of TEA buffer (pH 8.0), 0.6 μL of the purified thiol DNA, and 0.4 μL of Ellman's reagent. Let this reaction proceed for 10 min before taking a measurement.
11. Add an additional 540 μL of TEA buffer (pH 8.0) to the cuvet and take another measurement. The 260-nm measurements are used to calculate the DNA concentration, while the 412-nm measurements are used to calculate the free thiol concentration (*see* **Note 4**).

3.1.2. DNA Array Fabrication

1. Cover a 1.8 × 1.8 cm SF-10 slide with 1.5% Cytop solution.
2. Spin coat the slide at 500 rpm for 5 to 10 s, and then manually ramp the speed up to 5000 rpm and allow it to spin for at least 30 s (*see* **Note 5**).
3. Place the slides Cytop face up into a Pyrex® Petri dish, cover, and bake at 70°C for 50 min.
4. Remove the slides from the oven and place them in a different oven preheated to 190°C for 1 h. This temperature is above the boiling point of the CT-180 perfluorosolvent.
5. Place the Cytop-coated glass slides face up into a sample holder and cover with the mask containing the 700-μm holes. Mount the sample assembly into a vapor deposition chamber.
6. Vapor deposit 1 nm of chromium and then 45 nm of gold onto the Cytop-modified slides though the stainless steel mask. This will create gold patches on the hydrophobic glass surface (*see* **Note 6**).
7. Place the gold dot slide into an ethanolic MUAM solution for at least 4 h to form a well packed amine-terminated monolayer on the gold dots.
8. Remove the gold dot slide from the MUAM solution and rinse with absolute ethanol, then with water, and dry under a stream of nitrogen.
9. A PV830 Pneumatic Pico Pump (World Precision Instruments) is used to deliver 40- to 100-nL volumes of a freshly prepared SSMCC solution to the individual

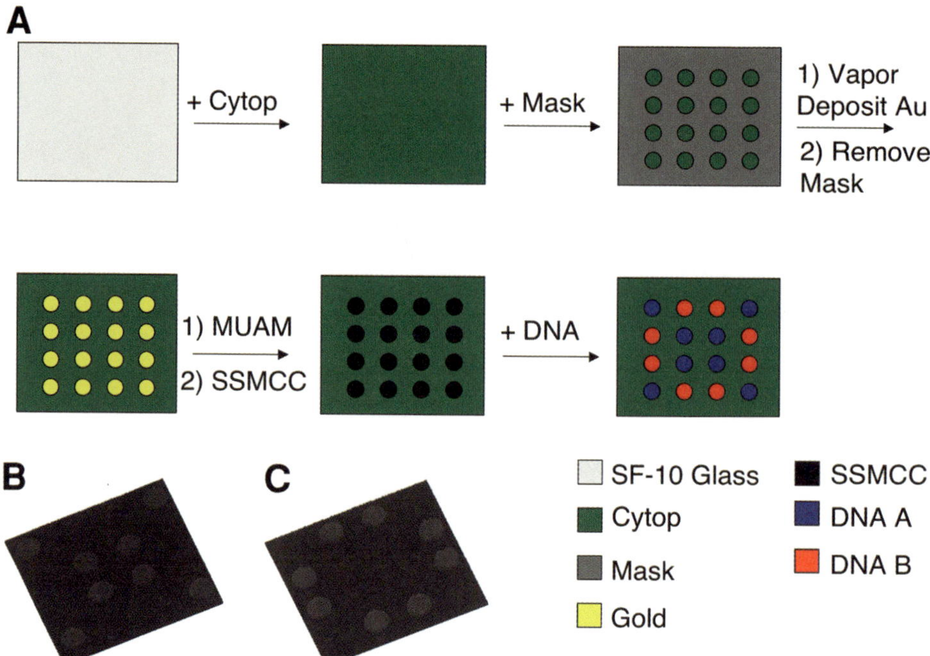

Fig. 1. **(A)** Schematic showing the DNA array fabrication process. **(B)** Surface plasmon resonance (SPR) difference image of DNA probe A binding to its complementary DNA sequence. The array is then denatured using 8 *M* urea to regenerate the single-stranded DNA array surface. **(C)** SPR difference image of DNA probe B binding to its complementary DNA sequence on the same array.

MUAM array elements within the Cytop-coated glass background. Allow the SSMCC to react with the surface for 20 min in order to form a maleimide surface that is thiol reactive.

10. Wash the slide with water and then dry under a stream of nitrogen. Using the pico pump, spot the thiol-modified DNA sequences onto the maleimide-terminated array elements to create an array of DNA molecules on the surface. Leave the thiol-modified DNA to react with the surface overnight.
11. Clean the surface with water to wash away the excess DNA solution, and dry under a stream of nitrogen. The DNA array should then be immediately mounted into the sample holder of the SPR imager or stored in a humidity chamber. An overview of the array fabrication process can be seen in **Fig. 1A**. **Figure 1B,C** shows SPR difference images for the sequence specific binding of complementary DNA sequences to a two-component DNA array. When the array is exposed to a DNA sequence complementary to DNA probe A, only those array elements are visible in the SPR difference image, indicating a hybridization/adsorption event (**Fig. 1B**). The hybridized DNA is then denatured with 8 *M* urea, and the same array is used

Fig. 2. Sample holder/flow cell assembly components.

to detect a DNA sequence complementary to probe B. When the array is exposed to the complementary sequence of probe B, only changes in the probe B elements are visible in the SPR difference image (**Fig. 1C**).

3.1.3. SPR Imaging Flow Cell Assembly

1. These instructions assume the use of an SPR imager from GWC Technologies. It is important that the three optical sides of the prism be thoroughly cleaned with methanol and lens paper. There should be no visible marks of any kind on the three faces of the prism. (*See* **Fig. 2** for an overview of the cell assembly components.)
2. Using tweezers, place the prepared array into the sample holder with the gold side facing down, using the four corners cut into the holder as a guide. Hold the array only by the corners to avoid damaging the functional array surface with the tweezers.
3. Clean the glass side of the gold slide using a cotton swab and methanol to remove any dust or smudges from the surface.
4. Without touching the surface, put one drop of index-matching fluid (Cargille, RI = 1.720) onto the center of the chip (*see* **Fig. 3**). Gently lower one of the optical sides of the prism on top of the chip at a slight angle to avoid bubble formation (*see* **Note 7**). Be sure to center the prism over the chip.
5. Place the prism wedge onto the top of the prism, and align the hole in the prism wedge with the tightening screw. If the hole in the prism wedge is not aligned with the tightening screw, gently move the wedge with tweezers until they are properly aligned. Hand-tighten the thumbscrew so that the chip and prism are held firmly in place.
6. Assemble the flow cell onto the chip surface, being careful to align the four screw holes of the flow cell with those in the sample holder. It is preferable that the inlet

Fig. 3. Prepared array placed in the sample holder with the gold side facing down. The four corners cut into the holder are used as a guide. A drop of index-matching fluid is then placed onto the center of the glass side of the array.

and outlet ports be positioned vertically when the assembled sample holder is placed into the SPR imager. Screw the flow cell into the sample holder while applying even pressure by incrementally tightening the screws diagonally from one another. Tighten the flow cell to the sample holder until the o-ring is visibly compressed (as viewed through the prism) against the gold slide.

7. Screw the inlet and outlet ports into the flow cell openings by hand. The fully assembled sample holder/flow cell assembly (shown in **Fig. 4**) can then be placed onto the SPR imager.

3.1.4. Mounting the Flow Cell and Angle Adjustment

1. Place the assembled flow cell onto the rotation stage so that the screw holes in the flow cell are aligned with the mounting holes in the rotation stage. Align the sample holder so that the light reflected from the gold surface will be directed toward the CCD camera. Fasten the flow cell to the rotation stage.
2. Connect the tubing from the pump to the lower inlet port of the flow cell. A section of tubing is connected to the top port and is used to deliver the waste solution to a beaker.
3. Place the end of the pump tubing into a small beaker of water, and turn the pump on to deliver water to the array surface.

Fig. 4. Fully assembled sample holder/flow cell assembly.

4. When the flow cell is completely filled with water, turn the pump off and transfer the pump tubing to an Eppendorf tube containing the buffer to be used in the SPR imaging experiment. Turn the pump on to deliver the buffer to the array surface.
5. Using the line profile function in the SPR imaging software, draw a line across the displayed image. This line on the image determines the points from which intensities are measured and displayed in the right-hand side of the window.
6. Turn the knob on the rotation stage until a minimum is observed in the line profile window, as determined from the pixel intensity values. This angle is known as the plasmon angle, and is the angle at which all incoming light is converted into surface plasmons. Turn the angle adjustment knob counterclockwise until the pixel value reaches a maximum. The angle is then adjusted to a point that is approximately one-third of the way from the minimum to the maximum pixel value. A background pixel intensity of approx 60–80 is a typical value, with respect to a maximum pixel intensity of 220.
7. This is a fixed-angle technique, and therefore it should not be readjusted during the course of the experiment.

3.1.5. SPR Imaging of DNA–Protein Binding

1. Fill the flow cell with response regulator protein-binding buffer and take an SPR image with an average of 30 capture frames (*see* **Note 8**). This will be the background image.
2. Flow complementary single-stranded DNA (500 n*M*) diluted in response regulator protein-binding buffer onto the array surface. Allow the DNA to hybridize to the

array elements for 15 min, and then flush the array with analyte-free buffer to remove nonspecifically adsorbed DNA from the background.

3. Take an SPR image (30 capture frames) of the array. Subtract the background image from this image to obtain an SPR difference image. The SPR difference image will show a change in percent reflectivity at those array elements in which hybridization has occurred. The double-stranded DNA array can then be used to detect double-stranded DNA binding proteins.
4. Flow response regulator protein-binding buffer over the array surface and take an SPR image. This image will serve as the background reference image.
5. Introduce a solution of transcription factor proteins, OmpR or VanR, diluted in response regulator protein-binding buffer to the array surface. These proteins can be imaged over a concentration range from 1 to 500 n*M*. The protein solution is allowed to sit on the surface for 15 min.
6. An SPR image is taken, and the background image (taken in **step 4**) is subtracted from this image. The resulting SPR difference image shows the specific binding of the transcription factor proteins to the surface. **Figure 5A** shows the specific binding of a 100 n*M* solution of OmpR to the double-stranded DNA array surface. Differential binding of the protein is observed for three of the probe sequences, with their relative binding affinities indicated by the relative signal intensity at each of the array elements (i.e., signal intensity increases as binding affinity increases). Changes in the SPR signal are not observed for the array elements designed to bind the protein VanR, and therefore they are not visible in the SPR difference image. The array was denatured with 8 *M* urea and then exposed to a 500 n*M* solution of VanR. **Figure 5B** shows the differential binding of the protein VanR to the VanR probe array elements. VanR shows the highest binding affinity to the Van H2 and Van R1 array elements, and significantly less binding to the Van H1 array elements. Slight nonspecific adsorption is observed at the OmpF1 array elements, whereas no change in the SPR signal is observed for the other array elements or to the PEG background.
7. Turn the polarizer 90 degrees, so that s-polarized light is striking the array surface. Take an SPR image, and then turn the polarizer back 90 degrees to return to p-polarized light. The value obtained for the s-polarized light will allow for the conversion of pixel intensity to change in percent reflectivity (*see* **Note 9**).
8. Wash the surface with 8 *M* urea, and allow it to remain on the surface for 15 min. This step will remove all of the complementary DNA and protein from the surface. Rinse the surface with water to wash away the urea. This process regenerates the single-stranded DNA array surface.
9. Repeat **steps 1–8** for each additional protein to be analyzed.

3.2. SPR Imaging of Peptide–Protein Interactions

3.2.1. Creation of 3D Silicon Wafer Masters and PDMS Microchannels

1. Pipet approx 4 to 5 mL of the negative photoresist onto the center of a silicon wafer and spin-coat at 5000 rpm for 60 s so that the wafer is evenly coated. Bake the wafer at 65°C for 5 min.

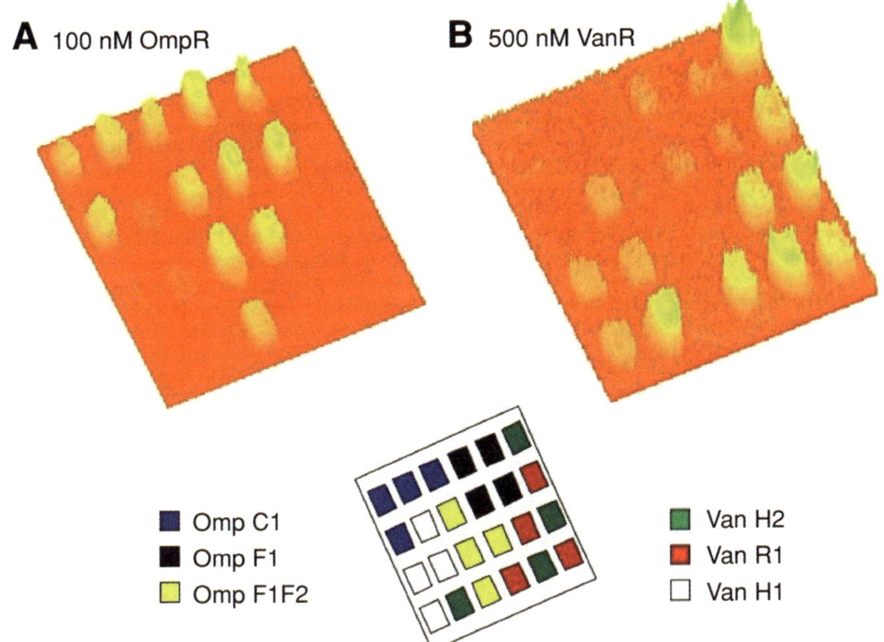

Fig. 5. Surface plasmon resonance (SPR) difference images of a six-component DNA array containing three sequences specific for OmpR and three specific for VanR. A double-stranded DNA array was first formed by exposing the array surface to a solution containing DNA complementary to all of the DNA probe sequences. **(A)** SPR difference image showing the binding of a 100 n*M* solution of the protein OmpR to the double-stranded DNA surface. Only the array elements specific for OmpR can be seen. **(B)** SPR difference image showing the binding of a 500 n*M* solution of the protein VanR to the double-stranded DNA array surface. Only the array elements specific for VanR show changes in the SPR difference image. Differences in binding intensity at each probe sequence are attributed to differences in their binding affinity to the specific protein. (Reprinted from **ref.** *8*, with permission of the American Chemical Society.)

2. Assemble the photoresist-covered silicon wafer and patterned chrome mask onto the sample stage of a UV aligner. Shine 365-nm UV light though the assembly for 40 s to transfer the pattern from the mask to the silicon wafer surface.
3. Bake the silicon wafer at 90°C for 5 min and then place the wafer in developer solution for 15 min at room temperature. This forms protruding shapes on the silicon wafer surface at regions that were previously exposed to the UV light in **step 2**.
4. Silanize the silicon master by storing it in a desiccator under vacuum for 2 h with a vial containing a few drops of tridecafluoro-1,1,2,2-tetrahydrooctyl-1-trichlorosilane. This last step is necessary to ensure the easy removal of the PDMS replicas from the 3D silicon master. These wafers can be reused for several months.

5. Mix the PDMS prepolymer with curing agent in a 10:1 ratio and pour it onto a 3D silicon wafer that has been placed into a plastic Petri dish. Be sure to pour enough PDMS onto the silicon master to cover the entire surface to a depth of 1 to 3 mm.
6. Allow the PDMS to de-gas for 20 min, until no air bubbles are visible, and then put the 3D silicon master and PDMS into an oven at 70°C for at least 1 h (*see* **Note 10**).
7. Take the PDMS and 3D silicon master from the oven and cut around the outside of the silicon master with a scalpel so that it can be removed from the Petri dish.
8. Carefully peel the PDMS microchannels from the surface of the silicon master and place them into a clean Petri dish.
9. Using a scalpel, cut out the desired microchannel so that it will fit onto the gold thin film (1.8 × 1.8 cm maximum).
10. Punch out reservoir holes at the ends of the PDMS microchannels so that target solution can be delivered through the microchannel (*see* **Note 11**). PDMS microchannels should be used only once and made just prior to use.

3.2.2. S-Peptide Array Fabrication

1. Soak a gold slide in an ethanolic solution of MUAM for at least 4 h to form a self-assembled monolayer of amine-terminated alkanethiols. Take the slide from the ethanolic solution, wash with ethanol, then with water, and dry under a stream of nitrogen.
2. Attach a set of microchannels featuring multiple parallel channels to the surface as a means to deliver reagents to the chemically modified gold surface.
3. Flow a solution of the bifunctional linker SPDP through the microchannels and allow it to react for 2 h. The NHS ester of the SPDP forms a covalent amide linkage with the MUAM monolayer, forming an active disulfide-terminated surface.
4. Wash the microchannels with water and then flow a 2 m*M* solution of either S peptide or modified S peptide down each individual microchannel. Let the thiol-disulfide immobilization reaction proceed overnight. **Figure 6A** shows a schematic of the patterned S-peptide variants on the array surface.

3.2.3. Flow Cell Assembly for Use With S-Peptide Array

1. Follow **Subheading 3.1.3., step 1**.
2. Mold a set of PDMS microchannels from the aluminum master containing a serpentine design (*see* **Note 12**).
3. Treat the PDMS microchannels with oxygen plasma for 10 s immediately prior to flow cell assembly.
4. Place a set of serpentine PDMS microchannels into a specially designed sample holder with prefabricated inlet and outlet ports that seal to the microchannels.
5. Place the gold slide onto the microfluidic channels face down, making sure to orient the immobilized lines of peptides perpendicular to the serpentine channels. SPR detection regions (300 μm × 670 μm) are formed where the immobilized peptide lines intersect with the PDMS microchannel. **Fig. 6b** shows a schematic of the serpentine PDMS microfluidics used to deliver analyte to the array surface throughout the SPR imaging experiment.

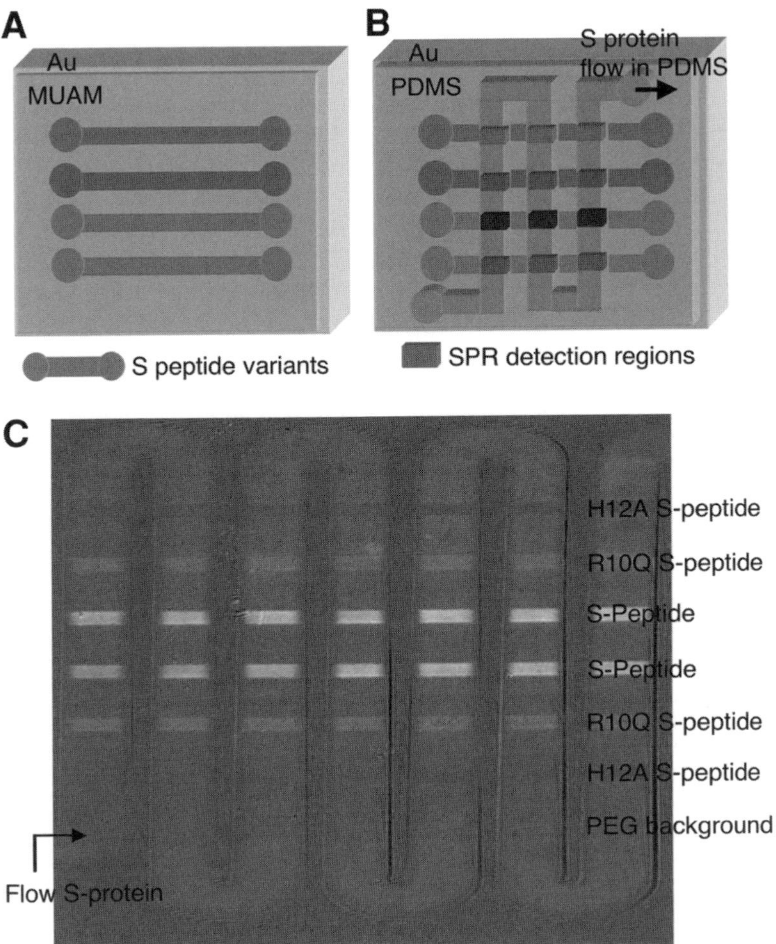

Fig. 6. **(A)** Schematic of S-peptide immobilization onto the chemically modified gold surface. **(B)** Schematic showing the flow direction of S protein through the polydimethylsiloxane microfluidics onto the peptide array. **(C)** Surface plasmon resonance difference image for a 75 n*M* solution of S protein binding to various S-peptide variants. Native S peptide shows the highest amount of binding to S protein, whereas the R10Q S-peptide variant containing a substitution of an amino acid at position 10 in the S-peptide sequence shows a much lower binding affinity. The H12A S-peptide variant shows no binding affinity to the S protein.

6. Without touching the surface, put one drop of index matching fluid (Cargille, RI = 1.720) onto the center of the back of the chip. Gently lower one of the optical sides of the prism onto the top of the chip at a slight angle to avoid bubble formation. Center the prism over the chip and tighten.

3.2.4. Mounting the PDMS Microfluidic Flow Cell and Angle Adjustment

1. Place the assembled flow cell onto the rotation stage so that the screw holes in the sample cell are aligned with the mounting holes in the rotation stage. Align the sample holder so that the light reflected from the gold surface will be directed towards the CCD camera. Fasten the flow cell to the rotation stage.
2. Connect the inlet port to a syringe pump for continuous sample delivery.
3. Flow PEG-NHS solution through the microchannnel and allow it to react for 1 h. The PEG-NHS will react with the surface amine groups in the regions surrounding the pattern of S peptides. This will result in a background that is resistant to the nonspecific adsorption of biomolecules.
4. The microchannel is then flushed with water to remove the solution of PEG-NHS.
5. Flow S-protein binding buffer through the PDMS microchannel using a syringe pump.
6. Using the line profile function in the SPR imaging software, draw a line across the displayed image. This line on the image determines the points from which intensities are measured and displayed in the right hand side of the window.
7. Turn the knob on the rotation stage until a minimum is observed in the line profile window, as determined from the pixel intensity values. This angle is known as the plasmon angle, and is the angle at which all incoming light is converted into surface plasmons. Turn the angle adjustment knob counterclockwise until the pixel value reaches a maximum. The angle is then adjusted to a point that is approximately one-third of the way from the minimum to the maximum pixel value. A background pixel intensity of approx 60–80 is typical.
8. Once the angle is adjusted, do not change it over the course of the experiment.

3.2.5. SPR Imaging of S Peptide–S Protein Equilibrium Binding

1. Flow a solution of S protein-binding buffer across the array surface through the PDMS microchannels using the syringe pump. Take an SPR image (30-frame average). This image will serve as the reference background image.
2. Flow a solution of S protein through the serpentine microchannel. Let the protein solution sit on the surface for 5 min to allow for equilibrium binding.
3. Take an SPR image with the protein still present in the microchannel. Subtract the background image (*see* **step 1**) from this image to obtain an SPR difference image for the affinity binding interaction. Changes in percent reflectivity are observed only where there is specific binding of S protein to the immobilized S peptide (*see* **Note 13**). **Figure 6C** shows an SPR difference image for the detection of S protein onto an array of S-peptide variants. Differential binding of the protein is clearly visible with a range of intensities observed at the various peptide array elements. Native S peptide shows the highest amount of binding to S protein (i.e., the peptide elements which appear to be the brightest in the image). The R10Q S-peptide variant containing a substitution of an amino acid at position 10 in the S-peptide sequence has a much lower binding affinity for the S protein. A negligible binding affinity of the S protein onto both the H12A S-peptide variant and the PEG background was observed (i.e., no changes in the SPR intensity).
4. Turn the polarizer 90 degrees, so that s-polarized light is striking the array surface. Take an SPR image, and then turn the polarizer back 90 degrees to return to

p-polarized light. The value obtained for the s-polarized light will allow for the conversion of pixel intensity to change in percent reflectivity.

5. Flush the array with buffer containing no analyte to regenerate the peptide array surface. The array can be used multiple times.

3.2.6. Real-Time SPR Imaging Measurements of S-Protein Binding

1. Fabricate the S-peptide array as outlined in **Subheading 3.2.2.**
2. Assemble the array onto the flow cell and mount onto the SPR imager as per **Subheadings 3.2.3.** and **3.2.4.**
3. Using custom-written software, a desired number of regions of interest (ROIs) on the array surface are defined. **Figure 7A** shows an SPR image with dotted lines indicating the boundaries of the ROIs on the chip. These ROIs correspond to different peptide array elements and PEG background correction regions.
4. Begin running the real-time data collection software prior to flowing the solution of S protein across the array surface. The collection macro calculates the change in average pixel intensity with respect to a corresponding reference value for each ROI and updates a graphical display in real-time. Typically, five frames are averaged with a time resolution of approx 1 s.
5. Flow analyte-free S-protein binding buffer over the array surface through the microfluidic channel to monitor the desorption curve of S protein from the S-peptide array surface.
6. These measurements can be repeated over a range of S-protein concentrations (i.e., 10 n*M* to 300 n*M*). From these measurements, values for the adsorption and desorption coefficients and binding affinity can be extracted.

4. Notes

1. Unless stated otherwise, all solutions should be prepared in water that has a resistance of 18.2 MΩcm. This standard is referred to as "water" in the text.
2. Commercially available DNA can be obtained with a wide variety of 5′ or 3′ modifications for surface-attachment chemistry. This method assumes the use of 5′ C6 S-S thiol-modified DNA or 3′ C3 S-S thiol-modified DNA.
3. The HPLC gradient used for DNA purification is as follows:
   a. Hold at 10% buffer B for the first 5 min.
   b. Gradually ramp to 70% buffer B over the next 45 min.
   c. Ramp to 100% buffer B in the next 2 min to wash the column.
   d. After 15 min, return to 10% buffer B.
4. The extinction coefficient to be used for the thiol concentration calculation is 13,600 L/(mol·cm).
5. The estimated thickness of the Cytop layer is less than 20 nm. In this particular application, it is important to achieve a Cytop layer as thin as possible to avoid losing SPR sensitivity.
6. Gold dot patterned slides should be immediately immersed and stored in an ethanolic solution of MUAM after vapor deposition. These arrays are also commercially available (GWC Technologies).
7. If a bubble forms in the index-matching fluid between the prism and array, carefully remove the slide from the prism using a pair of tweezers, taking care not to

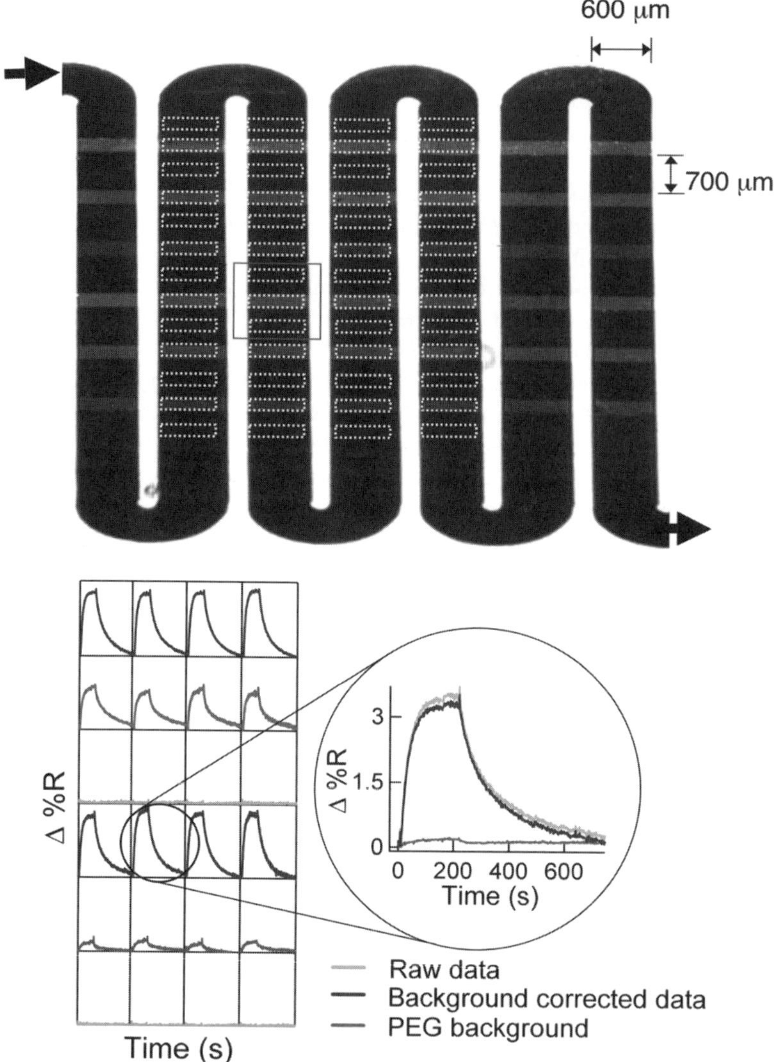

Fig. 7. **(A)** Representative raw surface plasmon resonance image of an S-peptide array with regions of interest (ROIs) drawn as dotted boxes on the array surface. The ROIs designate the position on the array where changes in percent reflectivity are to be measured as a function of time using a charge-coupled device camera. **(B)** Graphs showing the kinetics information obtained at each peptide array element when exposed to a solution of S protein. All of the curves were normalized with respect to adjacent polyethylene glycol regions. (Reprinted from **ref.** ***9***, with permission of the American Chemical Society.)

contaminate the array surface with the index-matching fluid. Using a cotton swab dipped in methanol, clean the glass side of the array to remove the index-matching fluid. Clean the prism and then remount the array and prism into the SPR sample holder.

8. An initial washing of the DNA probe surface with urea and water through the flow cell prior to beginning an experiment helps to clean the surface and ensure the best SPR imaging results.
9. Pixel intensity can be converted to change in percent reflectivity by the following equation:

$$\frac{0.85\,I}{S_{pol}} \times 100 = \Delta\%\ \text{Reflectivity}$$

where I is the SPR pixel intensity and S_{pol} is the pixel intensity measured using s-polarized light.

10. For both array fabrication and microfluidic flow cell preparation, it is preferable to cure the PDMS in the oven overnight.
11. When punching the reservoir holes in the PDMS, it is preferable to have the channel side facing up to avoid contaminating the channels with small PDMS particulates.
12. The serpentine microchannels fabricated from the aluminum master have a size of 670 μm width, 9.5 cm total length, 200 μm depth, and 400 μm spacing between folds, and are prepared in the same way as those fabricated on the silicon master (*see* **Subheading 3.2.1., steps 5–10**).
13. The solution of S protein must not be removed from the surface when taking an SPR measurement, because the S protein quickly disassociates from the immobilized S-peptide surface under nonequilibrium conditions.

References

1. Portefaix, J. M., Thebault, S., Bourgain-Guglielmetti, F., et al. (2000) Critical residues of epitopes recognized by several anti-p53 monoclonal antibodies correspond to key residues of p53 involved in interactions with the mdm2 protein. *J. Immunol. Methods* **244,** 17–28.
2. Zhu, H., Bilgin, M., Bangham, R., et al. (2001) Global analysis of protein activities using proteome chips. *Science* **293,** 2101–2105.
3. Zhu, H., Klemic, J. F., Chang, S., et al. (2000) Analysis of yeast protein kinases using protein chips. *Nat. Genet.* **26,** 283–289.
4. Shumaker-Parry, J. S., Zareie, M. H., Aebersold, R., and Campbell, C. T. (2004) Microspotting streptavidin and double-stranded DNA arrays on gold for high-throughput studies of protein-DNA interactions by surface plasmon resonance microscopy. *Anal. Chem.* **76,** 918–929.
5. Shumaker-Parry, J. S. and Campbell, C. T. (2004) Quantitative methods for spatially resolved adsorption/desorption measurements in real time by surface plasmon resonance microscopy. *Anal. Chem.* **76,** 907–917.
6. Garland, P. (1996) Optical evanescent wave methods for the study of biomolecular interactions. *Q. Rev. Biophys.* **29,** 91–117.

7. Hanken, D. G., Jordan, C. E., Frey, B. L., and Corn, R. M. (1996) Surface plasmon resonance measurements of ultrathin organic films at electrode surfaces, in *Electroanalytical Chemistry: A Series of Advances*, vol. 1 (Bard, A. J., ed.). Marcel Dekker, NY: pp. 141–225.
8. Smith, E. A., Erickson, M. G., Ulijasz, A. T., Weisblum, B., and Corn, R. M. (2002) Surface plasmon resonance imaging of transcription factor proteins: interactions of bacterial response regulators with DNA arrays on gold films. *Langmuir* **19,** 1486–1492.
9. Wegner, G. J., Wark, A. W., Lee, H. J., Codner, E., Saeki, T., Fang, S., et al. (2004) Real-time surface plasmon resonace imaging measurements for the multiplexed determination of protein adsorption/desorption kinetics and surface enzymatic reacations on peptide microarrays. *Anal. Chem.* **76,** 5677–5684.
10. Lee, H. J., Goodrich, T. T., and Corn, R. M. (2001) SPR imaging measurements of 1-D and 2-D microarrays created from microfluidic channels on gold thin films. *Anal. Chem.* **73,** 5525–5531.

8

Surface Plasmon Resonance Mass Spectrometry for Protein Analysis

Dobrin Nedelkov and Randall W. Nelson

Summary

The combination of surface plasmon pesonance (SPR) and mass spectrometry (MS) provides a unique methodology for studying proteins and their interactions. SPR is utilized to assess protein quantitative variations and the kinetic aspects of protein interactions, whereas MS complements the analysis by providing an exclusive look at the structural features of the interacting proteins via measurement of their mass. Thus, intrinsic protein structural modifications that go unregistered via the SPR detection can readily be assessed from the MS data. The purpose of this chapter is dissemination of the procedures and protocols for successful SPR-MS analysis. The individual steps of the complete SPR-MS process are illustrated via analysis of cardiac troponin I (cTnI).

Key Words: SPR; mass spectrometry; MALDI-TOF; protein interactions.

1. Introduction

Surface plasmon resonance (SPR) is a unique, label-free, direct-reading method of detection that utilizes interaction of light photons with free electrons (surface plasmons) on a gold surface to quantify the changes in concentration/amount of biomaterial on the surface *(1)*. The SPR phenomenon has been known for almost a century, but only in the last decade has it become accepted as a method for detection of biomolecular interactions, and been subsequently incorporated in a number of commercially available biosensors *(2)*. Today, SPR-based biosensors are mostly used for characterizing protein interactions under conditions where only one analyte of interest (typically in a pure form in solution) is targeted by an immobilized ligand on a single site on the sensor chip surface *(3–5)*. This experimental design is a result of the fact that SPR does not discriminate in the types of molecules it detects, as it registers only the total

From: *Methods in Molecular Biology, Vol. 328: New and Emerging Proteomic Techniques*
Edited by: D. Nedelkov and R. W. Nelson © Humana Press Inc., Totowa, NJ

amount of biomaterial retained on the surface. Hence, events such as multi-protein and nonspecific binding cannot be differentiated and can have a detrimental effect on the kinetic analysis. Although specificity in SPR is implied by the immobilized affinity ligand, a second specificity measure/dimension would certainly add to the approach. To improve upon the SPR measurements, in the mid-nineties we combined SPR with matrix-assisted laser desorption/ionization time-of-flight (MALDI-TOF) mass spectrometry (MS), in what has become known as SPR-MS, or biomolecular interaction analysis mass spectrometry (BIA/MS) ***(6,7)***. Because the SPR detection itself is nondestructive to the proteins being analyzed, the same proteins that are affinity-retrieved on the SPR chips can be further analyzed via MALDI-TOF or electrospray (ESI) mass spectrometry. Hence, MS validates the protein by looking at its intact mass, which is an intrinsic property of each protein. Furthermore, the MS analysis offers an additional insight into possible protein modifications that might exist as a result of point mutations or postexpression in vivo processing.

Even though a decade old, the SPR-MS approach is still not widely practiced. This is partly due to the fact that both SPR and MS are relatively novel technologies, and there are very few laboratories with expertise in both fields, which is essential for the realization of the combined approach. Furthermore, the interface between SPR and MS can be somewhat complicated, requiring specific skills and some practice for successful day-to-day operations. And finally, there are two schools of thought on how to practice SPR-MS. In the first, as promulgated by our laboratory, MS analysis is performed directly from the chip surface ***(8–19)***. In the second approach, the proteins are microrecovered from the chip surface and then MS analyzed from the elution solution ***(20–25)***. Described here are the protocols and the know-how for successful and reproducible integration of SPR and MS in the "from-the-chip" MS analysis approach.

2. Materials

2.1. Reagents

1. Human plasma sample: human blood was obtained from a single subject recruited within Intrinsic Bioprobes, Inc. (IBI), following a procedure approved by IBI's Institutional Review Board (IRB), and after signing of an informed consent form. Seventy-five microliters of blood were drawn under sterile conditions from a lancet-punctured finger with heparinized microcolumn (Drummond Scientific Co., Broomall, PA), mixed with 75 μL of HBS-EP buffer, and centrifuged for 30 s at 2500*g* to pellet the red blood cells. The supernatant (threefold-diluted plasma) was collected and used as is.
2. HBS-EP running buffer: 0.01 *M* HEPES (pH 7.4), 0.15 *M* NaCl, 0.005% (v/v) polysorbate 20, 3 m*M* ethylenediamine tetraacetic acid (EDTA).
3. Antibody coupling buffer: 10 m*M* acetate (pH 5.0).

4. Mouse monoclonal antibody to troponin I (Clone 8I-7, Spectral Diagnostics, Toronto, Canada).
5. Troponin I standard (HyTest Ltd, Turku, Finland, Cat. No. 8T53).
6. MALDI matrix: α-cyano-4-hydroxycinnamic acid (ACCA, Aldrich, Milwaukee, WI, Cat. No. 47,687-0), further processed by powder-flash re-crystallization from a low-heat saturated acetone solution of the original stock.

2.2. Surface Plasmon Resonance

Any SPR biosensor that utilizes a planar SPR chip can be utilized for the SPR analysis and protein retrieval. The dominant player in the SPR instruments field is Biacore AB, Uppsala, Sweden. Other new manufacturers of SPR biosensors include GWC Technologies, Inc., Madison, WI, and HTS Biosystems, East Hartford, CT. In the case of Biacore instruments, ligands (e.g., antibodies) are most commonly immobilized on the carboxymethyldextran surface of a CM5 Research Grade Sensor Chip (Biacore AB, cat. no. BR. 1000-14) using amine coupling kit chemicals (Biacore AB, cat. no. BR-1000-50).

2.3. Surface Plasmon Resonance–Mass Spectrometry Interface

In our laboratory, the interface between the SPR biosensor and the mass spectrometer consists of a chip cutter and matrix applicator. The chip cutter is made of a circular heated cutter-head that, when pressed against the plastic chip holder, excises a chip/plastic mount of a defined circular shape that fits into our MALDI mass spectrometer target *(8)*. The MALDI matrix applicator is an aerosol-spraying device also developed in our laboratory *(8)*. The device is essentially a pressurized microsprayer, consisting of an aspirating/sheath gas needle backed by approx 30 psi of compressed air. The air/matrix solution ratio can be adjusted to produce a fine mist of matrix solution that is aimed at the entire surface of the cutout chip.

2.4. Mass Spectrometry

MALDI-TOF mass spectrometry analysis from the chip surface can be performed on any commercially available mass spectrometer. In our laboratory, a custom-made MALDI-TOF mass spectrometer (Intrinsic Bioprobes, Inc.) is used for the SPR-MS analysis. The instrument consists of a linear translation stage/ion source capable of precise targeting of each of the flow cells under a focused laser spot. Ions generated during a 4-ns laser pulse (357 nm, nitrogen) are accelerated to a potential of 30 kV over a single-stage ion extraction source distance of approx 2 cm, before entering a 1.5-m field-free drift region. The ion signals are detected using a two-stage hybrid (channel plate/discrete dynode) electron multiplier. Time-of-flight spectra are produced by signal averaging of individual spectra from 50–100 laser pulses (using a 500-Mhz, 500-Ms/s digital transient recorder). Custom software is used in acquisition and analysis of the mass spectra. All protein spectra are obtained in the positive ion mode.

3. Methods

The individual steps of the entire SPR-MS process are illustrated by analysis of cardiac troponin I (cTnI). Cardiac troponin I is considered one of the most specific and sensitive markers of myocardial cell death. cTnI is proteolytically degraded in vivo ***(26,27)***, yielding multiple forms that circulate in plasma. Methods capable of simultaneous detection of these forms are needed, as some of these protein isoforms can potentially be used as a better predictor of myocardial infarction. To test the detection of truncated cTnI via SPR-MS, we set up two test tubes containing 100 μL of 10^{-3} mg/mL cTnI in HBS-EP buffer. One of the tubes also contained 10 μL of the human plasma. Both tubes were left overnight at room temperature with the intent of generating cTnI fragments in the tube containing the human plasma.

Mouse monoclonal antibody to troponin I was immobilized in both flow cells on a surface of a CM5 Research Grade Sensor Chip (*see* **Note 1**), using standard EDC/NHS coupling protocol and reagents supplied by Biacore. The antibody was diluted 100-fold in the acetate-coupling buffer. The SPR signals observed at the end of the immobilization process indicated successful immobilization of approx 85 fmoles of antibody in each flow cell (*see* **Note 2**). HBS-EP at a flow rate of 5 μL/min (*see* **Note 3**) was used as a running buffer in the SPR experiments. Following buffer equilibration in both flow cells, a single 50-μL aliquot of the control solution (10^{-3} mg/mL cTnI) was injected over FC1, resulting in an SPR response of 340 resonance units (RUs) (**Fig. 1**). The flow was then switched to FC2 and 50 μL of the cTnI-plasma solution was injected, resulting in an SPR response of 816 RUs. Immediately following the end of the second injection, the buffer flow was stopped (*see* **Note 4**), the chip removed from the biosensor, washed with three 200-μL aliquots of ultra-pure sterile water, cut from its plastic housing (*see* **Note 5**), and prepared for MALDI-TOF mass spectrometry by application of MALDI matrix (*see* **Note 6**) with the help of the matrix applicator device (*see* **Note 7**). **Figure 2** shows the mass spectra obtained from the surfaces of the two flow cells. Both mass spectra contain singly and doubly charged cTnI signals. However, the signals coming from flow cell 2 are shifted toward lower *m/z* values, indicating a truncation in the protein of approx 800 Da. The sample injected over FC2 was incubated overnight in the presence of a small aliquot of human plasma. Hence, the data verified the fact that cTnI is degraded upon release from myocardial tissue and entrance into the circulation. This example shows a rapid and effective way of assessing and measuring posttranslational modifications via the SPR-MS approach.

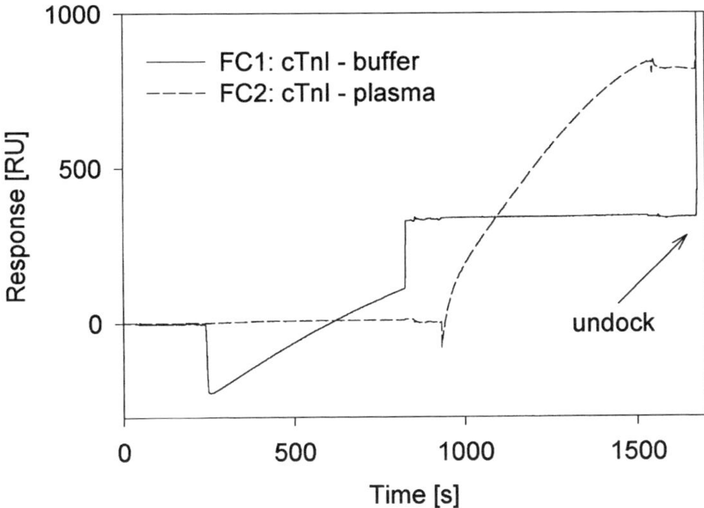

Fig. 1. Surface plasmon resonance sensorgrams showing the injection of 50-μL aliquots of 10^{-3} mg/mL cTnI with and without plasma over FC1 and FC2.

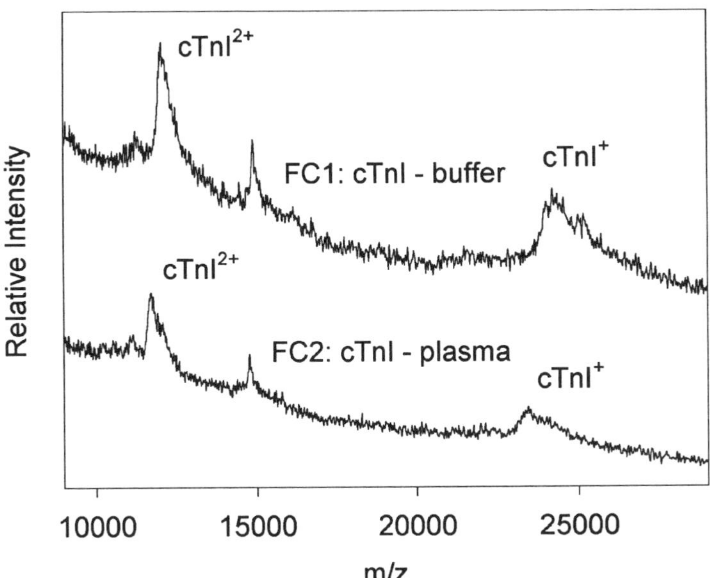

Fig. 2. Mass spectra taken from the surfaces of FC1 and FC2, showing the presence of multiply charged cTnI and truncated-cTnI ions, respectively (MW_{cTnI} = 23,916.3).

4. Notes

1. Prior to insertion into a Biacore SPR Biosensor, the sensor chip should be removed from the plastic housing cassette and washed with several aliquots of ultra-pure water, to remove any residual stabilizers that might interfere with the post-SPR-MS analysis. The chips should be then air dried, re-inserted into the plastic housing, and placed into the instrument using the standard docking command.
2. It is highly recommended that affinity ligands be immobilized in a high density on the chip surface. Although high ligand densities are disadvantageous to good kinetics data (i.e., re-binding of the analyte might complicate the kinetics analysis), they are very beneficial for the subsequent MS analysis because the amount of protein analyte captured from solution is increased. Because the intent of most SPR-MS analyses is affinity capture followed by MS, the generation of good kinetics data is somewhat secondary.
3. As another way to increase the amount of captured analyte, the flow rate of the sample solution over the ligand-derivatized flow cell surface should be decreased to approx 5 μL/min. This low flow rate promotes mass transfer across the flow cell, which in turn increases the amount of protein analyte captured on the chip. Otherwise, at high flow rates typically used for SPR kinetics experiments (60–100 μL/min), low-concentration analytes will not be captured in amounts conducive to downstream MS analysis.
4. The time between the end of the SPR analysis and the subsequent preparation steps for MS should be minimized in order to avoid any protein losses in this interim period. Hence, following the end of the sample injection step, the SPR response should be promptly recorded, the flow of the running buffer stopped, and the chip removed from the biosensor (the fastest way to undock the chip is via the root control software of the biosensor, i.e., the OS9 terminal configured for connection to the instrument service port). Once outside the biosensor, the chip should be removed from the plastic cassette and washed with several aliquots of ultra-pure water to remove the residual buffer components (such as detergents), which might cause interference in the mass spectra.
5. In all of our past SPR-MS work, we have used a chip cutter with circular heated cutter head (Intrinsic Bioprobes Inc., Tempe, AZ) for excising a chip/plastic mount of a defined circular shape that fits into an appropriately configured MALDI mass spectrometer target *(8)*. However, most commercially available mass spectrometers can accommodate the large footprint of the entire chip housing, as long as the MALDI target is appropriately machined/milled to accommodate the entire chip. Because of the different thickness of the chip (as compared to a standard MALDI target), little drops of protein standards should also be placed on the corners of the chip to aid in the mass spectra calibration.
6. For the MS analysis from the chips, ACCA has been our matrix of choice. ACCA is a better energy-absorbing matrix than sinapic acid (SA) and, consequently, requires less laser power. Lower laser power means that more spectra can be obtained from a single spot, and fast burning through the matrix/sample layer is avoided; it should be noted that the analyte is captured in a very thin layer on the

surface of the chip, and aggressive application of the laser (at high frequency) can rapidly deplete the sample spot. Re-application of more matrix (following initial application and MS analysis) does not generally yield better signals, and results in decreased S/N ratio.

7. The application of MALDI matrix is the most critical step in the entire SPR-MS process. The matrix mist generated by the matrix applicator should moisten but not completely wet the chip surface (i.e., the tiny matrix droplets should stay as individual drops on the surface, and not be connected into one large liquid drop). The matrix droplets will desorb the proteins from their respective capturing affinity ligand and, upon rapid drying, the matrix/protein mixture will be re-deposited on the same area from where the proteins were originally captured in the SPR analysis.

Acknowledgments

This publication was supported in part by grant number 5 R44 CA099117-03 from the National Institutes of Health. Its contents are solely the responsibility of the authors and do not necessarily represent the official views of the National Institutes of Health.

References

1. Homola, J., Yee, S. S., and Gauglitz, G. (1999) Surface plasmon resonance sensors: review. *Sensors Actuat. B* **54,** 3–15.
2. Liedberg, B., Nylander, C., and Lundstrom, I. (1995) Biosensing with surface-plasmon resonance—how it all started. *Biosens. Bioelectron.* **10,** R1–R9.
3. Cooper, M. A. (2003) Label-free screening of bio-molecular interactions. *Anal. Bioanal. Chem.* **377,** 834–842.
4. Homola, J. (2003) Present and future of surface plasmon resonance biosensors. *Anal Bioanal Chem.* **377,** 528–839.
5. Karlsson, R. (2004) SPR for molecular interaction analysis: a review of emerging application areas. *J. Mol. Recognit.* **17,** 151–161.
6. Krone, J. R., Nelson, R. W., Dogruel, D., Williams, P., and Granzow, R. (1996) Interfacing mass spectrometric immunoassays with BIA. *BIAJournal* **3,** 16–17.
7. Nelson, R. W., Krone, J. R., and Jansson, O. (1997) Surface plasmon resonance biomolecular interaction analysis mass spectrometry. 1. Chip-based analysis. *Anal. Chem.* **69,** 4363–4368.
8. Nedelkov, D. and Nelson, R. W. (2000) Practical considerations in BIA/MS: optimizing the biosensor-mass spectrometry interface. *J. Mol. Recognit.* **13,** 140–145.
9. Nedelkov, D. and Nelson, R. W. (2000) Exploring the limit of detection in biomolecular interaction analysis mass spectrometry (BIA/MS): detection of attomole amounts of native proteins present in complex biological mixtures. *Anal. Chim. Acta* **423,** 1–7.
10. Nedelkov, D. and Nelson, R. W. (2001) Analysis of human urine protein biomarkers via biomolecular interaction analysis mass spectrometry. *Am. J. Kidney Dis.* **38,** 481–487.
11. Nedelkov, D. and Nelson, R. W. (2001) Analysis of native proteins from biological fluids by biomolecular interaction analysis mass spectrometry (BIA/MS): exploring

the limit of detection, identification of non-specific binding and detection of multi-protein complexes. *Biosens. Bioelectron.* **16,** 1071–1078.

12. Nedelkov, D. and Nelson, R. W. (2001) Delineation of in vivo assembled multi-protein complexes via biomolecular interaction analysis mass spectrometry. *Proteomics* **1,** 1441–1446.
13. Nedelkov, D. and Nelson, R. W. (2003) Design and use of multi-affinity surfaces in biomolecular interaction analysis-mass spectrometry (BIA/MS): a step toward the design of SPR/MS arrays. *J. Mol. Recognit.* **16,** 15–19.
14. Nedelkov, D. and Nelson, R. W. (2003) Delineating protein-protein interactions via biomolecular interaction analysis-mass spectrometry. *J. Mol. Recognit.* **16,** 9–14.
15. Nedelkov, D. and Nelson, R. W. (2003) Surface plasmon resonance mass spectrometry: recent progress and outlooks. *Trends Biotechnol.* **21,** 301–305.
16. Nedelkov, D. and Nelson, R. W. (2003) Detection of Staphylococcal enterotoxin B via biomolecular interaction analysis mass spectrometry. *Appl Environ Microbiol.* **69,** 5212–5215.
17. Nedelkov, D., Nelson, R. W., Kiernan, U. A., Niederkofler, E. E., and Tubbs, K. A. (2003) Detection of bound and free IGF-1 and IGF-2 in human plasma via biomolecular interaction analysis mass spectrometry. *FEBS Lett.* **536,** 130–134.
18. Nelson, R. W., Nedelkov, D., and Tubbs, K. A. (2000) Biomolecular interaction analysis mass spectrometry. BIA/MS can detect and characterize proteins in complex biological fluids at the low- to subfemtomole level. *Anal. Chem.* **72,** 404A–411A.
19. Nelson, R. W., Nedelkov, D., and Tubbs, K. A. (2000) Biosensor chip mass spectrometry: a chip-based proteomics approach. *Electrophoresis* **21,** 1155–1163.
20. Gilligan, J. J., Schuck, P., and Yergey, A. L. (2002) Mass spectrometry after capture and small-volume elution of analyte from a surface plasmon resonance biosensor. *Anal. Chem.* **74,** 2041–2047.
21. Kikuchi, J., Furukawa, Y., and Hayashi, N. (2003) Identification of novel p53-binding proteins by biomolecular interaction analysis combined with tandem mass spectrometry. *Mol. Biotechnol.* **23,** 203–212.
22. Natsume, T., Taoka, M., Manki, H., Kume, S., Isobe, T., and Mikoshiba, K. (2002) Rapid analysis of protein interactions: On-chip micropurification of recombinant protein expressed in *Esherichia coli*. *Proteomics* **2,** 1247–1253.
23. Lopez, F., Pichereaux, C., Burlet-Schiltz, O., Pradayrol, L., Monsarrat, B., and Esteve, J. P. (2003) Improved sensitivity of biomolecular interaction analysis mass spectrometry for the identification of interacting molecules. *Proteomics* **3,** 402–412.
24. Zhukov, A., Schurenberg, M., Jansson, O., Areskoug, D., and Buijs, J. (2004) Integration of surface plasmon resonance with mass spectrometry: automated ligand fishing and sample preparation for MALDI MS using a Biacore 3000 biosensor. *J. Biomol. Tech.* **15,** 112–119.
25. Borch, J. and Roepstorff, P. (2004) Screening for enzyme inhibitors by surface plasmon resonance combined with mass spectrometry. *Anal. Chem.* **76,** 5243–5248.

26. Shi, Q., Ling, M., Zhang, X., et al. (1999) Degradation of cardiac troponin I in serum complicates comparisons of cardiac troponin I assays. *Clin. Chem.* **45,** 1018–1025.
27. Katrukha, A. G., Bereznikova, A. V., Filatov, V. L., et al. (1998) Degradation of cardiac troponin I: implication for reliable immunodetection. *Clin. Chem.* **44,** 2433–2440.

9

High-Throughput Affinity Mass Spectrometry

Urban A. Kiernan, Dobrin Nedelkov, Eric E. Niederkofler, Kemmons A. Tubbs, and Randall W. Nelson

Summary

Affinity mass spectrometry (AMS) is a proteomics approach for selectively isolating target protein(s) from complex biological fluids for mass spectrometric analysis. The resulting high-content mass spectrometry (MS) data show the unique MS protein signatures (wild-type, posttranslationally modified, as well as genetically modified forms of the protein target) that are present within a biological sample. Information regarding such protein diversity is normally lost in classical proteomic or immunoassay analyses. This chapter presents a step-by-step description of high-throughput AMS in the population proteomic screening of the human plasma protein cystatin C.

Key Words: Affinity purification; mass spectrometry; human plasma proteins; high throughput.

1. Introduction

Clinical proteomics has evolved into a global protein biomarker discovery program, the goal of which is to mass-spectrometrically identify novel proteins that are useful for disease diagnosis/prognosis as well as therapeutic intervention. Current classical proteomics initiatives have already identified numerous candidate biomarkers; however, validation of these targets for clinical application is problematic as a result of issues of reproducibility, throughput, and data content. Such validation studies are often reserved for conventional immunoassay platforms (enzyme-linked immunosorbent assay [ELISA], radioimmunoassay [RIA], etc.), but their indirect detection methods are blind to protein population variations, which are a result of genetic and/or posttranslational modifications (truncations, oxidation, phosphorylation, etc.). It has now become apparent that such variants need to be accounted for in any analysis, as subtle protein modifications can be the cause as well as the manifestation of disease.

From: *Methods in Molecular Biology, Vol. 328: New and Emerging Proteomic Techniques*
Edited by: D. Nedelkov and R. W. Nelson © Humana Press Inc., Totowa, NJ

Therefore, second-generation proteomics technologies, which selectively target a single protein as well as its associated variants, have assumed great importance. A novel approach to targeted proteomics analysis is affinity mass spectrometry (AMS), a hybrid of micro-affinity capture with matrix-assisted laser desorption/ionization time-of-flight (MALDI-TOF) MS detection *(1)*. This approach allows for reproducible high-throughput analyses, producing high-content data that readily differentiate multi-affinity forms of the same target protein *(2,3)*. Presented here is an in-depth description of the mechanics involved in the application of high-throughput AMS technologies, targeting the human plasma protein cystatin C (CYSC) for a routine population proteomics analysis.

2. Materials

1. Human plasma from a population of 96 individuals. The Li-heparin plasma samples used in this study were purchased from a commercial tissue bank (ProMedDx, LLC, Norton, MA).
2. A 96-sample format robotic workstation (Multimek 96, Beckman Coulter, Fullerton, CA).
3. One box of 96 activated underivatized carboxyl affinity pipets (Intrinsic Bioprobes, Inc., Tempe, AZ) (*see* **Note 1**).
4. Antibody solution: anti-CYSC polyclonal antibody (A0451, DakoCytomation, Carpinteria, CA). Solution stored at 4°C.
5. Ligand-coupling buffer: 10 m*M* $NaCH_3CO_2$ (pH 5.1). Solution prepared fresh.
6. Blocking reagent: 1 *M* ethanolamine (ETA, 39,813-6, Sigma-Aldrich, St. Louis, MO) (pH 8.5). Solution stored at 4°C.
7. Hydrochloric acid: 60 m*M*. Solution stored at room temperature.
8. Reconditioning buffer: 10 m*M* HEPES-buffered saline (pH 7.4), 150 m*M* NaCl. Solution stored at 4°C.
9. Dilution and rinse buffers: 10 m*M* HEPES-buffered saline (pH 7.4), 150 m*M* NaCl, 3 m*M* ethylenediamine tetraacetic acid (EDTA), and 0.005% Surfactant P20 (HBS). Solution stored at 4°C.
10. Double-distilled water.
11. Organic rinse: 20% (v/v) acetonitrile with 2 *M* ammonium acetate (pH 7.5). Solution stored at 4°C.
12. Normalization rinse solution: 10 m*M* *n*-octylglucoside (NOG, 10281772, Roche Diagnostics, Mannheim, Germany). Solution stored at 4°C.
13. MALDI matrix: 10 g/L sinapic acid (SA, 85429, Fluka, Milwaukee, WI) in 33% (v/v) acetonitrile and 0.4 % (v/v) trifluoroacetic acid (TFA) (*see* **Note 2**).
14. Calibration solution: 1×10^{-3} mg/mL equine cytochrome C (MW = 12,360.2; 9007-43-6, Sigma-Aldrich, St. Louis, MO) in double-distilled water. Solution stored at 4°C.
15. A 96-well formatted MALDI target (Intrinsic Bioprobes, Inc., Tempe, AZ).
16. Linear Autoflex MALDI-TOF mass spectrometer (Bruker Daltonics, Billerica, MA).

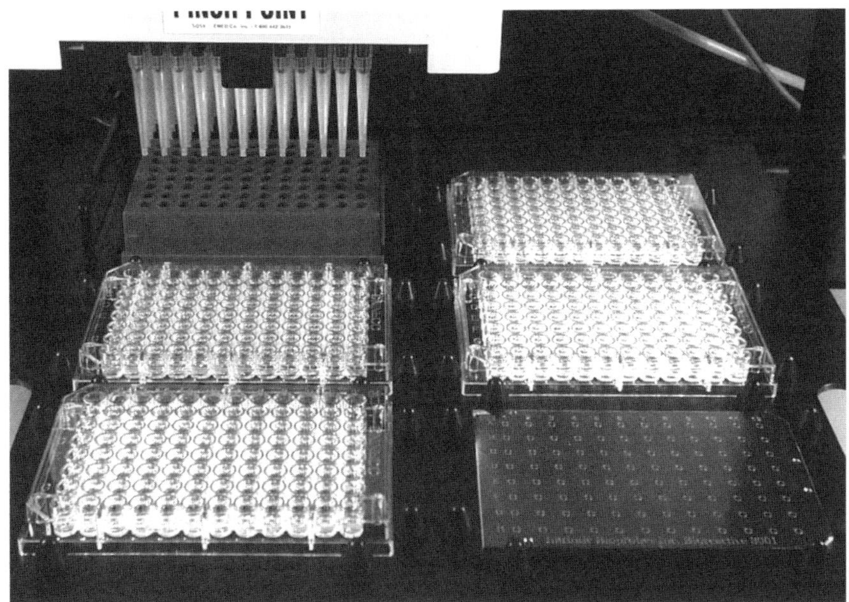

Fig. 1. High-throughput robotics platform outfitted with affinity pipets for automated sample preparation and protein capture. The footprint consists of a six-stage work area that can be used for automated affinity pipet derivatization, sample analysis, and analyte elution.

17. Proteome Analyzer Software (Intrinsic Bioprobes Inc., Tempe, AZ) on a Windows-compatible computer station.

3. Methods

Affinity mass spectrometry revolves around micro-affinity pipet technology, produced by Intrinsic Bioprobes, Inc. (IBI, Tempe, AZ). These affinity pipets, when derivatized with an affinity ligand, are capable of selectively retrieving and enriching protein target(s) from a variety of complex biological fluids *(4–6)* and have been successfully applied to proteins at sub-μg/L concentrations *(3,7)*. The functional affinity pipet format is easily converted into a high-throughput platform by integrating a parallel processing robotics workstation *(2,8–10)*. This automated system can be utilized for affinity pipets preparation (derivatization with affinity ligand) as well as sample analysis (**Fig. 1**).

3.1. Derivatization of Affinity Pipets With Antibody

1. All materials for the derivatization of the affinity pipets are made ready: the anti-CYSC antibody solution is diluted 1:100 in the ligand-coupling buffer and dispensed into individual wells of a 96-well microplate in 100-μL aliquots; the ETA,

hydrochloric acid, and reconditioning buffer solutions are also dispensed into 96-well microplates, but at 200 μL per well.

2. The underivatized affinity pipets are loaded onto the robotic workstation along with reagent microplates (**Fig. 1**).
3. The loaded affinity pipets first address the diluted antibody solution. The solution is repetitively passed through the affinity pipets 400 times (50 μL per aspiration), during which, antibody is covalently bound to the activated solid support within each affinity pipet *(11)*.
4. After coupling, residual binding sites are blocked by flowing ETA solution through the affinity pipets 50 times (100 μL per aspiration).
5. Blocked affinity pipets are then rinsed with a mild acid solution. During the derivatization process, antibodies may aggregate together, and the acid rinse is used to liberate noncovalently attached antibody. The 60 m*M* hydrochloric acid is flowed through the affinity pipets 30 times (100 μL per aspiration). This step is repeated in a second microplate containing fresh reagent.
6. The immobilized antibody is then reconditioned by exposure to a buffer solution at neutral pH. The reconditioning buffer is also flowed through the affinity pipets 30 times (100 μL per aspiration). This step is also repeated in a second microplate of fresh reagent.
7. Once reconditioned, the derivatized affinity pipets are ready for use. However, they can be stored wet in reconditioning buffer at 4°C for periods up to 3 mo (*see* **Note 3**).

3.2. Sample Preparation and Affinity Purification

1. Reagents and materials necessary for sample affinity purification are prepared: human plasma samples are aliquotted (100 μL per sample) into individual wells of a 96-deep-well microplate (*see* **Note 4**). Samples are then diluted to 1 mL with 900 μL of dilution buffer. Rinsing buffers and double-distilled water are aliquotted into 96-well microplates (200 μL per well). Buffers, water, and samples are warmed to room temperature prior to the initiation of analysis (*see* **Note 5**).
2. The derivatized affinity pipets are loaded into the robotic workstation along with samples and rinse microplates (*see* **Note 6**).
3. Affinity pipets are first rinsed with HBS (100 μL flowed through the affinity pipets 10 times) to remove any debris or contaminants that the affinity pipets may possibly have come in contact with since they were prepared.
4. Rinsed affinity pipets are then immersed into the diluted plasma samples. The plasmas are flowed through the affinity pipets 150 times at 100 μL per aspiration.
5. After probing the diluted plasma samples, the affinity pipets are enriched with the targeted protein. Nonspecifically retained proteins may accompany the target protein and are removed using a series of rinses. The first rinse is HBS, which is flowed through the affinity pipets 10 times at 100 μL per aspiration.
6. The second rinse is double-distilled water. This too is flowed through the pipets 10 times at 100 μL per aspiration.
7. The next step is the organic rinse. The organic rinse mixture is flowed through the affinity pipets 10 times at 100 μL per aspiration (*see* **Note 7**).

8. The final rinse is double-distilled water. This rinse is split between two microplates and is used to purge the affinity pipets of any residual salt. The water is flowed through the affinity pipets 20 times (100 μL per aspiration).

3.3. Protein Elution

1. Reagents required for the elution of the affinity-retained proteins are prepared. This includes distributing 150-μL aliquots of the 10 m*M* NOG solution into the wells of a microplate, as well as pouring the 50 mL of bulk MALDI matrix in a basin for parallel aspiration (*see* **Note 8**). At this point, a MALDI target is positioned within the robotic system for eluant deposition (**Fig. 1**).
2. The elution protocol is initiated on the robotic workstation (*see* **Note 9**). The affinity pipets are first treated with the normalization rinse. This involves a single aspiration (100 μL per affinity pipet) of the NOG solution. Affinity pipets are blown out and blotted to remove any residual NOG solution (the robotic elution method will pause to allow for manual blotting of the affinity pipets).
3. Next, the affinity pipets undergo parallel aspiration (6 μL per affinity pipet) of the matrix solution.
4. The affinity pipets, laden with matrix solution, are then positioned above the MALDI target, where droplets of the matrix/protein eluates are formed and then deposited directly onto the MALDI target surface (**Fig. 2**). The target is imprinted with protein samples in the same 8 × 12 array format as the one in which the proteins were captured from the samples (*see* **Note 10**).
5. After eluates deposition, the used affinity pipets can be discarded while the imprinted samples are air-dried.
6. The drying MALDI target is then manually spotted with a calibration standard on each of its four corners. A 4-μL spot containing a 1:1 mixture of calibration solution with MALDI matrix is used for each calibration point.

3.4. MALDI-TOF MS Analysis

1. Once the imprinted protein matrix samples are crystallized, the MALDI target is inserted into the mass spectrometer for analysis.
2. Intact protein MALDI-TOF MS analysis is performed on a linear Bruker Autoflex instrument. The system is operated in positive ion, delayed-extraction mode with a 20.00 kV full accelerating potential. Draw-out pulses of 1.350 kV and a 390 ns delay are used for this analysis (*see* **Note 11**).
3. Resulting mass spectral data can be calibrated, viewed, and contrasted using the Proteome Analyzer Software (*see* **Note 12**). The sample spectra are externally calibrated with the mass spectra acquired from the cytochrome C standards. A CYSC protein profile was successfully produced for each sample. An overlay of all 96 mass spectra is shown in **Fig. 3**. These target protein profiles showed the presence of wild-type CYSC (MW = 13,343) as well as an oxidized variant (MW = 13,359) (*see* **Note 13**). N-terminally truncated variants of CYSC—desS- (MW = 13,272) and desSSP- (MW = 13,072)—are also consistently observed. This basic profile is highly conserved among the majority of samples, as shown in **Fig. 4** upper trace; however, a

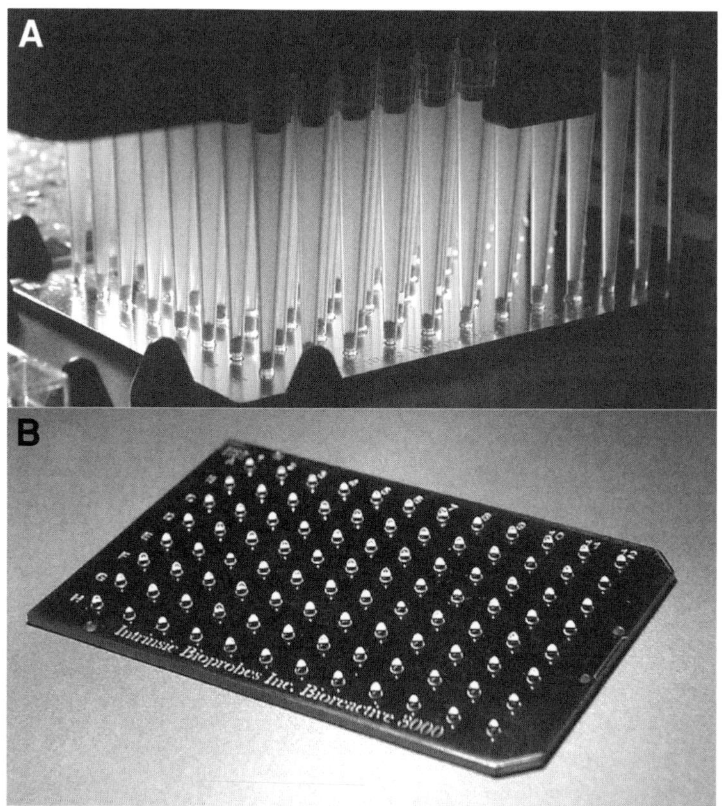

Fig. 2. Auto-deposition of affinity-purified analyte onto a 96-well formatted MALDI-target. **(A)** The eluate, a MALDI matrix/affinity purified protein mix, is stamped in parallel onto the target. **(B)** Results in an 8 × 12 array of protein samples that, when dry, are ready for mass spectrometric analysis.

few contained an increased number of truncated CYSC variants. The variants observed in **Fig. 4** middle trace show a progressive breakdown of CYSC as the protein is serially degraded from the N-terminus. These truncated forms include: desSSPG- (MW = 13,015), desSSPGK- (MW = 12,887), and desSSPGKPPR- (MW = 12,536). Eventually, CYSC degradation from both termini is indicated by the appearance of desSSPGKPPR- & -DA (MW = 12,350) and desSSPGKPPRLV- & -DA (MW = 12,138) truncated variants. A genetically mutated form of CYSC is also observed in the sample population (**Fig. 4** bottom trace) and is readily identifiable form the wild-type form by an approx +30-Da mass shift.

4. Notes

1. AMS affinity pipets are available in various chemical formats. Activated carboxyl affinity pipets were found to work best for plasma CYSC analysis and were utilized

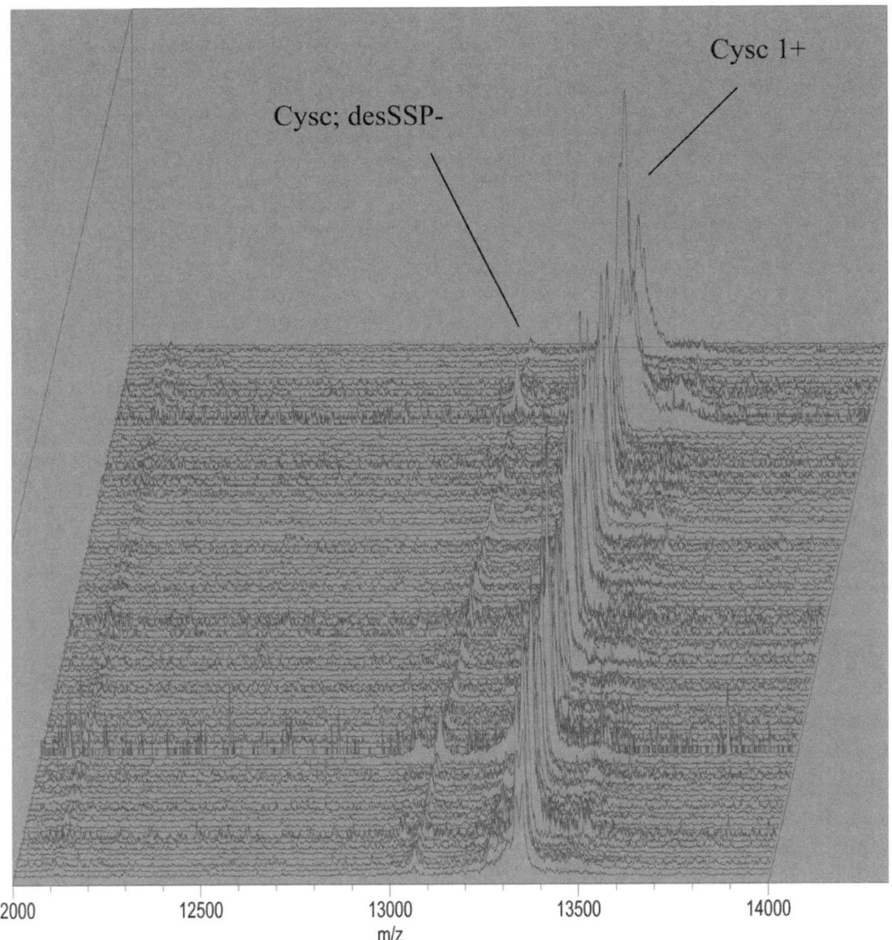

Fig. 3. Mass-spectrometric data for all 96 human samples. Cystatin C was detected from all samples, with the majority sharing the same basic protein profile.

in this example. The coupling protocol and reagents described are specific for the carboxyl affinity pipet format. Other affinity pipet chemistries may have enhanced affinity capture when used with other target proteins and in other biological fluids.

2. For high-throughput automated elution, a 50-mL matrix is prepared fresh.
3. Even though stability tests show only a minimal loss of activity when derivatized affinity pipets are stored wet at 4°C for a 3-mo period, we recommend immediate use for optimal results.
4. The plasma samples are thawed and then spun for 5 min at 3,000*g* to pellet any debris that may result from storage prior to being aliquotted. Moreover, sample pretreatment steps as well as internal reference standard doping may also be required for target and quantitative analysis *(7)*, but are not needed in this example.

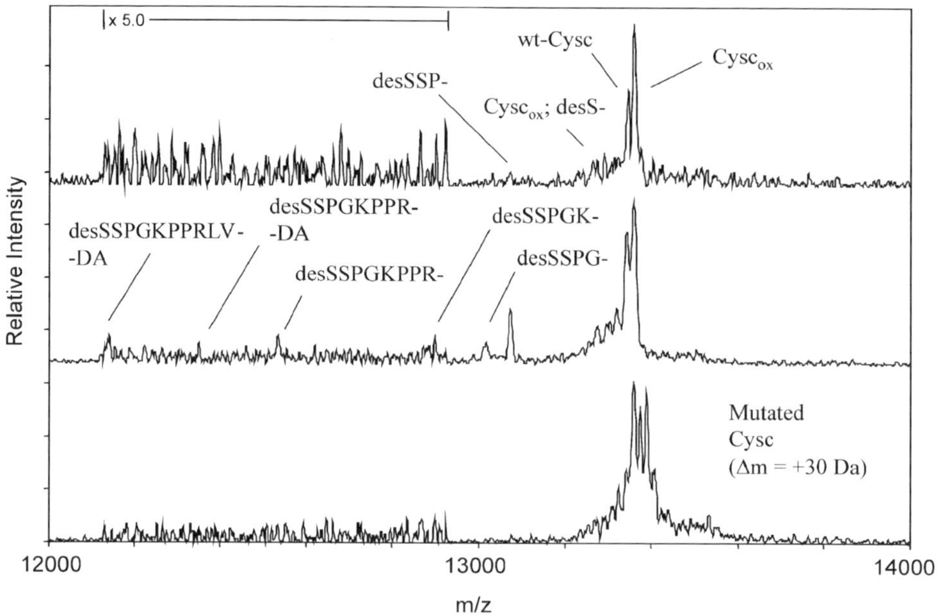

Fig. 4. Representative individual mass spectra from the targeted population proteomics analysis for Cystatin C. The mass spectrometric analysis reveals that CYSC is a heterogeneous mixture of wild-type CYSC and multiple variants. Upper trace: mass spectrometry (MS) trace that is representative of the majority of the samples observed in this high-throughput analysis. Wild-type, oxidized, and a truncated form of CYSC are readily observed. Middle trace: MS trace from an individual whose CYSC profile showed an increase in the number of truncation variants. Bottom trace: CYSC profile from an individual who is heterozygous for a CYSC point mutation, which results in a Δm = approx +30 Da.

5. Ensure that ample time is given to allow samples and reagents to warm to room temperature. Affinity capture is temperature dependent, and variations can lead to inconsistent results.
6. Prior to loading stored affinity pipets into the robotic workstation, manually remove any residual reconditioning buffer that may be contained within them. The system is unable to purge the affinity pipets of any initial volume of fluid. Such residual fluid may aspirate into the head mechanism and contaminate the robotic system.
7. Because the components of the organic rinse mixture are not highly miscible, the stock rinse is agitated and aliquotted just prior to use in the analysis. At that point, the robotic method applied will pause to allow for the insertion of the organic mix microplate into the rinse series.
8. Matrix is aspirated from a large basin that supports all 96 affinity pipets in a single aspiration step. Aspirating matrix from a microplate results in poor sample crystal formation and weak ion signals due to matrix precipitation within the individual

wells. Once the stock matrix solution is poured into the basin, let stand 5 min prior to elution, to allow any matrix sediment to settle.

9. The elution protocol is series of scripted macros that were written specifically for the AMS affinity pipet elution process.
10. The MALDI targets used are hydrophilic/hydrophobic contrasted, allowing for optimum sample spot concentration for higher-quantity MS data and more efficient data acquisition *(8)*.
11. The MALDI-TOF MS instrument setting used was determined to provide optimum sensitivity and resolution of proteins that are in the mass-to-charge (m/z) range of 11,500 to 13,500 thousand. Such settings must be adjusted for other protein targets that are not in this m/z range.
12. MALDI-TOF MS data are technically observed as m/z values; however, the +1 ion can be directly converted to molecular mass units (Da). For convenience and clarity, all data presented here are in molecular mass units.
13. Externally calibrated mass spectra have observed masses that range from 13,339 to 13,344 and 13,358 to 13,362, respectively. These values translate into a parent protein mass accuracy of 300 ppm. When used in conjunction with the specificity of affinity retrieval, such mass accuracies allow for variant identification with a high degree of certainty.

Acknowledgments

This project has been funded in part with federal funds from the National Institute of Environmental and Health Sciences, under Contract No. N44ES-35511. This publication was also supported in part by grant number 5 R44 CA099117-03 from the National Cancer Institute, National Institutes of Health. Its contents are solely the responsibility of the authors and do not necessarily represent the official views of the National Institutes of Health.

References

1. Nelson, R. W., Krone, J. R., Bieber, A. L., and Williams, P. (1995) Mass-spectrometric immunoassay. *Anal. Chem.* **67,** 1153–1158.
2. Kiernan, U. A., Tubbs, K. A., Gruber, K., et al. (2002) High-throughput protein characterization using mass spectrometric immunoassay. *Anal. Biochem.* **301,** 49–56.
3. Kiernan, U. A., Tubbs, K. A., Nedelkov, D., Niederkofler, E. E., McConnell, E., and Nelson, R. W. (2003) Comparative urine protein phenotyping using mass spectrometric immunoassay. *J. Proteome Res.* **2,** 191–197.
4. Tubbs, K. A., Nedelkov, D., and Nelson, R. W. (2001) Detection and quantification of beta-2-microglobulin using mass spectrometric immunoassay. *Anal Biochem.* **289,** 26–35.
5. Kiernan, U. A., Tubbs, K. A., Nedelkov, D., Niederkofler, E. E., and Nelson, R. W. (2002) Comparative phenotypic analyses of human plasma and urinary retinol binding protein using mass spectrometric immunoassay. *Biochem. Biophys. Res. Commun.* **297,** 401–405.

6. Kiernan, U. A., Nedelkov, D., Tubbs, K. A., Niederkofler, E. E., and Nelson, R. W. (2004) Proteomic characterization of novel serum amyloid P component variants from human plasma and urine. *Proteomics* **4,** 1825–1829.
7. Nelson, R. W., Nedelkov, D., Tubbs, K. A., and Kiernan, U. A. (2004) Quantitative mass spectrometric immunoasay of insulin like growth factor 1. *J. Proteome Res.* **3,** 851–855.
8. Niederkofler, E. E., Tubbs, K. A., Gruber, K., et al. (2001) Determination of beta-2 microglobulin levels in plasma using a high-throughput mass spectrometric immunoassay system. *Anal. Chem.* **73,** 3294–3299.
9. Kiernan, U. A., Nedelkov, D., Tubbs, K. A., Niederkofler, E. E., and Nelson, R. W. (2002) High-throughput analysis of human plasma proteins. *Am. Biotech. Lab.* **20,** 26–28.
10. Nedelkov, D., Tubbs, K. A., Niederkofler, E. E., Kiernan, U. A., and Nelson, R. W. (2004) High-throughput comprehensive analysis of human plasma proteins: a step toward population proteomics. *Anal. Chem.* **76,** 1733–1737.
11. Niederkofler, E. E., Tubbs, K. A., Kiernan, U. A., Nedelkov, D., and Nelson, R. W. (2003) Novel mass spectrometric immunoassays for the rapid structural characterization of plasma apolipoproteins. *J. Lipid Res.* **44,** 630–639.

10

Isotope-Coded Affinity Tags for Protein Quantification

Christopher M. Colangelo and Kenneth R. Williams

Summary

An important goal in proteomics is to compare the relative amounts of different proteins in biological samples and to try to correlate these differences with changes in physiological state. The isotope-coded affinity tag technique pioneered in Aebersold's laboratory takes advantage of differential tagging of cysteine residues in proteins with stable isotopes to significantly reduce the complexity of peptide mixtures and increase the number of sequences that are identified in a single tandem mass spectrometry experiment. In this approach, two samples are isotopically labeled (one heavy, one light) through a reactive group that specifically binds to cysteine residues; the samples are combined, separated with chromatography, and analyzed by mass spectrometry. The results are then database searched and a list of hundreds of proteins and their heavy:light ratio is obtained.

Key Words: Isotope labeling; proteomics; liquid chromatography tandem mass spectrometry; affinity labeling; ICAT.

1. Introduction

Isotope-coded affinity tag (ICAT) analysis profiles the relative amounts of cysteine-containing peptides derived from tryptic digestion of protein extracts. Recognizing that only a single tryptic peptide is needed to quantify the expression of the corresponding parent protein, the ICAT reagent was designed to affinity isolate and quantify, via the use of a stable isotope, the relative concentrations of cysteine-containing tryptic peptides obtained from digestion of control vs experimental samples *(1)*. Hence, the newest ICAT reagent from Applied Biosystems (ABI) (**Fig. 1**) has a thiol-specific reactive group adjacent to an alkyl linker, which contains either nine [^{12}C] or nine [^{13}C] atoms, resulting in a mass difference of 9 Da between the control and the corresponding experimental version of the same tryptic peptide. A nice feature of the ICAT approach is that the in vitro incorporation of a stable isotope into one of the two samples being compared obviates the need to analyze by mass spectrometry the control and experimental

From: *Methods in Molecular Biology, Vol. 328: New and Emerging Proteomic Techniques*
Edited by: D. Nedelkov and R. W. Nelson © Humana Press Inc., Totowa, NJ

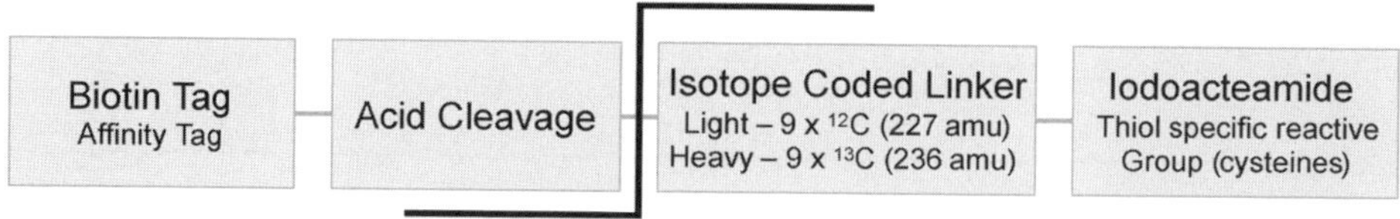

Fig. 1. Cartoon diagram of cleavable isotope-coded affinity tag reagent (Applied Biosystems).

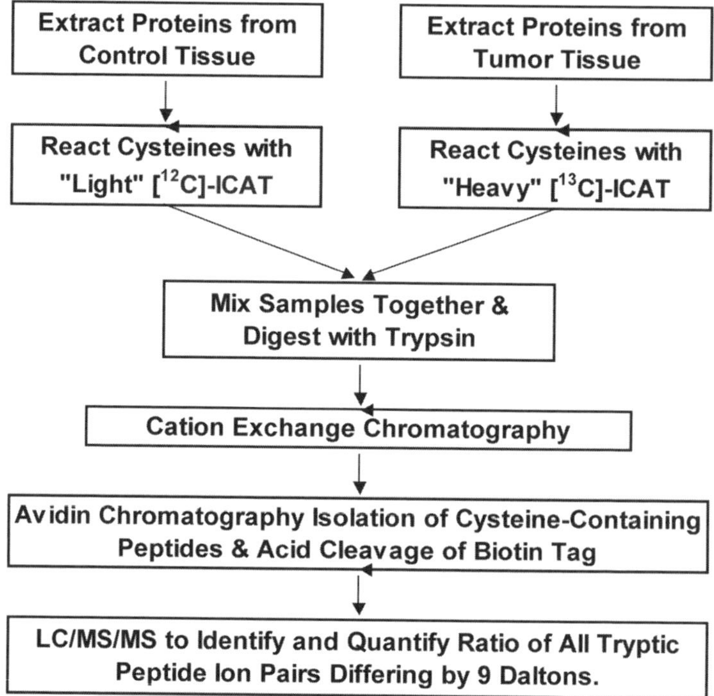

Fig. 2. Flow chart depicting isotope-coded affinity tag/mass spectrometry-based protein profiling.

samples separately. The alkyl linker in the ICAT reagent is connected to a (cleavable) biotin group, which allows rapid affinity isolation of cysteine-containing tryptic peptides. Although a tryptic digestion of a whole-cell human protein extract might produce 550,000 peptides, less than 100,000 of these might be expected to contain cysteine. Based on a search of the Swiss Database, less than 5% of human proteins lack cysteine and would be missed. As depicted in **Fig. 2**, following derivatization of the control protein extract with the [^{12}C]-ICAT reagent and of the experimental protein extract with the [^{13}C]-ICAT reagent, the pooled samples are subjected to trypsin digestion followed by cation exchange (CEX) chromatography. Typically, a whole-cell or tissue-protein extract would be divided into 36 cation exchange fractions, with each of them being subjected to avidin chromatography isolation of cysteine-containing tryptic peptides followed by liquid chromatography (LC)-tandem mass spectrometry (MS/MS) analysis to identify ICAT peptide pairs and quantify the relative [^{12}C]/[^{13}C] ratios. ICAT-derivatized peptide pairs that differ by exactly 9 Da are identified and quantified by the Applied Biosystems ProICAT software. ProICAT can perform modified database searches by extracting and using only data obtained on cysteine-containing peptides, thus significantly reducing search and data analysis times. ProICAT uses a three-dimensional LC/MS reconstruct algorithm to locate and accurately determine experimental:control (heavy:light) peak ratios in complex proteomic samples (*see* **Note 1**).

The resulting ICAT data, which are analogous to those obtained via the use of two different fluorescent dyes in DNA microarray analysis of mRNA or DIGE analysis of protein expression, provide the corresponding ratio for the level of expression of the parent protein in the control vs experimental sample. Currently, the largest number of proteins profiled by this approach from a single sample was 491 proteins contained in microsomal fractions of native and in vitro differentiated human myeloid leukemia cells ***(2)***. In order to perform high-throughput proteomic profiling using ICAT technology, it is essential to automate as many steps as possible. The standard ICAT procedure from Applied Biosystems requires *manual* syringe-based purification steps on both cation-exchange and avidin cartridges. To address these challenges, we use an Applied Biosytems Vision workstation to automate *both* these steps. By using the Vision workstation, we are able to automate both the cation exchange high-performance liquid chromatography (HPLC) and avidin cartridge chromatography steps. This greatly enhances our ability to process samples, to maximize sample throughput on the QSTAR XL mass spectrometer, and to substantially reduce the possibility of errors associated with manual syringe-based purification (*see* **Note 2**).

2. Materials

1. Vial of both C12 (light) and C13 (heavy) ABI Cleavable ICAT reagent (cat. no. 4339038).
2. ABI cleaving reagents A and B (supplied with ABI Cleavable ICAT reagent, cat. no. 4339038).
3. 100 µg of control and experimental protein sample.
4. High-resolution cation-exchange column (PolySulfoethyl A Column, 5 µm 300 Å bead, from PolyLC, Inc., 2.1 × 200 mm, cat. no. 202SE0503).
5. Sequencing-grade trypsin (100 µg, Promega cat. no. V5280).

6. Electrospray ionization mass spectrometer with tandem MS capability (e.g., ABI QSTAR XL mass spectrometer).
7. Nanoflow chromatography system (e.g., LC Packings Ultimate).
8. Nanoflow reversed-phase column (Atlantis Nanoease C18 Column from Waters, 100 μm × 150 mm, 3 μm bead, cat. no. 186002209).
9. Nanoflow reversed-phase trapping column (Atlantis Nanoease Trap column from Waters, 0.18 × 23.5 mm, cat. no. 186002574).
10. HPLC system with fraction collection (e.g., ABI Vision Workstation).
11. Denaturing buffer: 25 m*M* Tris, 0.1% sodium dodecyl sulfate (SDS) (pH 8.5).
12. Reducing reagent: 50 m*M* Tris(2-carboxyethyl)phosphine hydrochloride (TCEP).
13. CEX loading buffer: 10 m*M* KH_2PO_4, 25% acetonitrile (pH 3.0).
14. CEX clean buffer: 10 m*M* KH_2PO_4, 25% acetonitrile, 1 *M* KCl (pH 3.0).
15. CEX neutralizing buffer: 100 m*M* sodium phosphate (pH 10.0).
16. CEX storage buffer: 10 m*M* KH_2PO_4, 25% acetonitrile, 0.02% NaN_3 (pH 3.0).
17. Affinity loading buffer: 2X phosphate-buffered saline (PBS) (20 m*M* phosphate buffer, 150 m*M* NaCl), pH 7.2.
18. Affinity wash 2 buffer: 50 m*M* ammonium bicarbonate, 20% methanol (pH 8.3).
19. Affinity elute buffer: 30% acetonitrile, 0.4% trifluoroacetic acid (TFA).
20. Affinity storage buffer: 2X PBS (pH 7.2), 0.02% NaN_3.
21. Avidin Cartidges (Applied Biosystems, cat. no. 4326694).
22. Cartidge Holder (Applied Biosystems, cat. no. 4326688).
23. ABI ProICAT software on Windows-compatible workstation.

3. Methods

3.1. Denaturing and Reducing of Protein Extracts

1. Add 80 μL of denaturing buffer to two tubes containing 100 μg each of protein. If working with a concentrated sample solution, add denaturing buffer to bring volume up to 80 μL) (*see* **Notes 3–5**).
2. Add 2 μL of reducing reagent to each tube.
3. Vortex to mix, then spin (7000*g* for approx 15 s).
4. Place tubes in a boiling water bath for 10 min.
5. Vortex to mix, then spin for 1 to 2 min to cool tubes.

3.2. Labeling With Cleavable ICAT Reagents

1. Take one vial of both light and heavy cleavable ICAT reagent from freezer and wrap with aluminum foil. The cleavable ICAT reagent is light sensitive, so be sure to minimize exposure to light (*see* **Note 6**).
2. Wait 15 min to allow the ICAT reagents to reach room temperature before using.
3. Spin reagents to bring all powder to bottom of each vial.
4. Add 20 μL of acetonitrile to each reagent vial.
5. Vortex to mix, then spin.
6. Transfer contents of one of the protein vials to a vial of light ICAT reagent.
7. Transfer contents of the other protein vial to a vial of heavy ICAT reagent.

Table 1
Cation Exchange (CEX) Fractions for Isotope-Coded Affinity Tag Samples

Sample description	No. proteins expected	No. CEX fractions collected and subjected to LC-MS/MS
Small protein complexes	<10	1
Larger protein complexes	<200	12
Whole-cell extracts	>200	36

LC-MS/MS, liquid chromatography-tandem mass spectrometry.

8. Vortex to mix, then spin.
9. Incubate for 2 h at 37°C.
10. Vortex to mix, then spin.
11. Mix light-labeled sample with the heavy-labeled sample (this is your 1:1 sample, total volume = 200 μL).

3.3. Trypsin Digestion Procedure

1. Reconstitute a vial of trypsin (100 μg) with 200 μL of Milli-Q water.
2. Vortex to mix, then spin.
3. Add trypsin solution to the empty, light-ICAT-labeled sample vial.
4. Transfer trypsin solution from the light-ICAT-labeled sample vial to the heavy-ICAT sample vial that contains the 1:1 sample from **Subheading 3.2., step 11**.
5. Incubate 12 to 16 h at 37°C.
6. Vortex to mix, then spin.

3.4. Cation Exchange Chromatography

1. Depending on the number of proteins expected in the sample (<10, <200, or >200), refer to **Table 1** to determine the number of fractions collected during the cation-exchange step. Typically, for whole-cell extracts we collect 36 CEX fractions over the course of the gradient (*see* **Note 7**).
2. Take 400 μL of trypsin-digested ICAT-labeled peptides and dilute to 1 mL by adding approx 600 μL CEX loading buffer.
3. Check the pH of the sample to be sure it is less then 3.0, and if necessary add 1 *M* phosphoric acid to lower pH below 3.0.
4. Using the HPLC system, pre-equilibrate a 2.1 mm × 200 mm PolySulfoethyl A column with CEX loading buffer at a flow rate of 150 μL/min.
5. Load entire sample on column and run a linear gradient of CEX clean buffer from 0 to 98% B in 90 min at a flow rate of 150 μL/min. The gradient is designed to spread out the elution of doubly charged peptides, with these peptides usually eluting starting at about 50 min into the run until approx 65 min, after which triply charged peptides begin to elute.
6. Collect 200-μL fractions and store at 4°C if the avidin chromatography will be carried out immediately; otherwise, store fractions at –20°C.

7. Clean the column using 10 mL of CEX clean buffer, and if the column will not be immediately used again, wash and store the column with 10 mL of CEX storage buffer.

3.5. Avidin Chromatography

1. For each CEX fraction, a parallel avidin chromatography column will have to be run to isolate the Cys-containing tryptic peptides.
2. Take each CEX fraction and add 100 μL CEX neutralizing buffer to bring the pH to approx 7.0.
3. Dilute fraction 4:1 by adding 400 μL affinity loading buffer, and then check that the pH is approx 7.0.
4. Using the HPLC system, pre-equilibrate column with affinity loading buffer at a flow rate of 500 μL/min.
5. Load entire fraction onto the affinity column and then wash with 6 mL of affinity loading buffer to remove excess salt.
6. Wash nonspecifically bound peptides off the column with 6 mL of 1X PBS (50% affinity loading buffer, 50% H_2O) and follow with 6 mL of affinity wash 2 buffer.
7. Next remove wash buffer from sample by running 1 mL Milli-Q water over the column.
8. Elute Cys-containing tryptic peptides with 1.5 mL of affinity elute buffer and collect this eluent.
9. Clean the column with 2 mL of affinity elute buffer. If the column will be stored, wash the column with 6 mL of affinity storage buffer and store at 4°C.

3.6. Cleavage of ICAT-Labeled, Cys-Containing Tryptic Peptides

1. Dry down affinity-eluted fractions from the Avidin column in Speedvac overnight, without heat.
2. In a clean tube, prepare final cleaving reagent by combining 95 μL of cleaving reagent A and 5 μL of cleaving reagent B (from the ABI cleavable ICAT reagent kit).
3. Vortex to mix, then spin.
4. Transfer 45 μL of freshly prepared cleaving reagent to each sample tube.
5. Incubate for 2 h at 37°C.
6. Dry down samples in Speedvac.
7. Dissolve each sample in 8 μL of 0.1% formic acid.

3.7. LC-MS/MS Analysis

1. For LC-MS/MS analysis, separate the avidin-purified samples on a reverse-phase HPLC that is in front of and in line with a tandem MS-compatible mass spectrometer.
2. Inject 8 μL of sample onto a nanoflow reverse-phase trapping column at 20 μL/min in 0.1% TFA, 2% acetonitrile.
3. Place the nanoflow trap column inline with a 100 μm × 15 cm Waters Atlantis C18 column equilibrated with 0.1% TFA, 2% acetonitrile at a flow rate of 500 nL/min

Table 2
Typical Results From Two Human Cell Lines

	NB4/Monocytes	NB4/Neutrophil
Identified human proteins (two or more peptides/protein)	122	88
number of above proteins with ratios above/below threefold difference	40	10
number of human proteins identified with only one peptide	1554	615

and elute individual peptides with a 60-min gradient using 0.1% TFA, 5% isopropanol, 85% acetonitrile.

4. Using a data-dependent method, set up MS acquisition to carry out an MS scan for 0.5 s from 400 to 1200 amu, and to select the two most intense peaks with charge state 2 to 4 for an MS/MS product ion acquisition from 50 to 1800 amu with a 2-s accumulation time.
5. After nanoflow LC-MS/MS, import the data into ProICAT software to automatically quantify and identify differentially expressed proteins. The results of the ProICAT data are stored in a Microsoft Access database for future data mining.

4. Notes

1. Quantitation accuracy for the entire experiment is approx ±20–25% of the actual value, with a linear range of detection within a 10-fold difference between the control and treated samples. Differences in level of expression between individual proteins in the control vs experimental samples that exceed 10-fold are likely to be missed, as only the tryptic peptide from the protein that is expressed at the higher concentration is likely to be detected and it is not generally feasible to manually review all "singleton" LC/MS peaks.
2. **Table 2** provides an example of the overall number of proteins that might be expected to be identified and quantified using the ICAT technology to profile 100 μg crude protein extract from each of two human cell lines grown in culture and when 36 CEX fractions are collected and individually subjected to LC/MS.
3. Sample preparation is key to optimal ICAT reagent labeling. Ideally, samples should be free of contaminants (acid, salt, and detergents) and lyophilized. If this is not possible, samples should be concentrated to approx 5 mg/mL and a volume of 20 μL. Several methods, such as acetone precipitation, gel filtration chromatography/spin columns, micro-dialysis, or ultracentrifugation, can be used to concentrate and remove contaminants from protein extract samples prior to labeling. Additionally, low pH will hinder the iodoacetamide reaction; thus, it is imperative to keep the pH of the sample between 8.0 and 9.0. Excess detergents can interfere with the tryptic digest. Although we typically use SDS detergent, 6 *M* urea can be used instead, provided that the sample is diluted to <2 *M* urea prior to trypsin digestion.

Also, some laboratories have used octoglucoside, radioimmunoprecipitation (RIPA) buffer, and NP40, but these detergents need to be kept under 0.1% final concentration. Finally the overall salt concentration of the reaction needs to be kept below 40 m*M*.

4. Increasing the amount of protein above 100 µg/sample should allow the detection of less-abundant proteins; however, this will require the use of one vial of ABI ICAT reagent per 100 µg of protein labeled.
5. To ensure that the amount of extracted protein is carefully matched at 100 µg for each control or treated sample, we use hydrolysis/amino acid analysis (http://keck.med.yale.edu/prochem/aaa.htm) of an approx 1–5 µg aliquot of each sample to accurately determine the protein concentration prior to labeling. Other laboratories have used the Bradford assay to determine protein concentration, but our experience is that this assay has a ±20–25% measurement error, as compared to the ±5–10% measurement accuracy of amino acid analysis.
6. The ICAT reagents have a shelf life of 1 yr at –20°C.
7. The maximum loading capacity of the 2.1 mm × 200 mm polysulfoethyl A strong cation exchange column is approx 5 mg, but for optimum chromatography we recommend limiting the loading to 1 mg total protein.

Acknowledgments

This project has been funded in part with federal funds from National Heart, Lung and Blood Institute/National Institutes of Health (NIH) contract N01-HV-28186 and National Institute on Drug Abuse/NIH grant 1 P30 DA018343-01.

References

1. Gygi, S. P., Rist, B., Gerber, S. A., Turecek, F., Gelb, M. H., and Aebersold, R. (1999) Quantitative analysis of complex protein mixtures using isotope coded affinity tags. *Nat. Biotechnol.* **17,** 994–999.
2. Han, D. K., Eng, J., Zhou, H., and Aebersold, R. (2001) Quantitative profiling of differentiation-induced microsomal proteins using isotope-coded affinity tags and mass spectrometry. *Nat Biotechnol.* **19,** 946–951.

11

Proteomic Analysis by Multidimensional Protein Identification Technology

Laurence Florens and Michael P. Washburn

Summary

Multidimensional chromatography coupled to mass spectrometry is an emerging technique for the analysis of complex protein mixtures. One approach in this general category, multidimensional protein identification technology (MudPIT), couples biphasic or triphasic microcapillary columns to high-performance liquid chromatography, tandem mass spectrometry, and database searching. The integration of each of these components is critical to the implementation of MudPIT in a laboratory. MudPIT can be used for the analysis of complex peptide mixtures generated from biofluids, tissues, cells, organelles, or protein complexes. The information described in this chapter will provide researchers with details for sample preparation, column assembly, and chromatography parameters for complex peptide mixture analysis.

Key Words: Mass spectrometry; multidimensional chromatography; proteome; proteomics; database searching; MudPIT; informatics.

1. Introduction

The integration of multidimensional chromatography and mass spectrometry as a proteomic tool has developed into a mature technology. Numerous two-dimensional (2D) chromatographic methodologies have been developed in order to resolve peptide mixtures prior to analysis by mass spectrometry (reviewed in **ref.** ***1***). One technique, multidimensional protein identification technology (MudPIT), has proven itself a powerful method to study highly complex protein samples. In MudPIT, a bi- or triphasic microcapillary column packed with reversed-phase (RP) and strong cation exchange (SCX) high-performance liquid chromatography (HPLC)-grade materials is loaded with a complex peptide mixture generated from a biological sample ***(2–5)***. The microcapillary column is interfaced with a quaternary HPLC pump coupled to a tandem mass spectrometer,

From: *Methods in Molecular Biology, Vol. 328: New and Emerging Proteomic Techniques*
Edited by: D. Nedelkov and R. W. Nelson © Humana Press Inc., Totowa, NJ

and acts as the ion source. Peptides are directly eluted off of the microcapillary column, ionized, and analyzed in the tandem mass spectrometer, which is capable of fragmenting peptides in a predictable fashion, which allows for the computational determination of the peptide sequence by database-searching algorithms like SEQUEST ***(6)***. Utilizing MudPIT, researchers are able to measure protein levels of whole proteomes ***(7–9)***, differential protein expression in response to variable growth conditions ***(10)***, membrane proteins ***(11)***, and large multi-protein complexes ***(12,13)***.

2. Materials

2.1. Sample Preparation

1. Benzonase (Sigma, St. Louis, MO).
2. Trichloroacetic acid (TCA).
3. Digestion buffer: 100 m*M* Tris-HCl (pH 8.5) (stored at 4°C), 8 *M* urea (added fresh).
4. Tris(2-carboxylethyl)-phosphine hydrochloride (TCEP) (1 *M* stock, stored at –20°C, diluted 1/10) (Pierce, Rockford, IL).
5. Iodoacetamide (IAM), 0.5 *M* stock (Sigma, St. Louis, MO).
6. Endoproteinase LysC (Roche Applied Science, Indianapolis, IN).
7. Calcium chloride, 500 m*M* stock.
8. Trypsin, modified sequencing grade: 0.1 µg/µL stock in water at –20°C (Roche Applied Science, Indianapolis, IN).
9. 90% formic acid.
10. Elastase stock solution: 10 µg/µL in 10 m*M* Tris-HCl (pH 8.5), stored at –20°C (Calbiochem, San Diego, CA).
11. Subtilisin A stock solution: 10 µg/µL in 10 m*M* Tris-HCl (pH 8.5), stored at –20°C (Calbiochem, San Diego, CA).
12. Proteinase K stock solution: 15 µg/µL in water, stored at –20°C (Roche Applied Science, Indianapolis, IN).

2.2. Microcapillary Column Construction and Sample Loading

1. 100 µm i.d. × 365 µm o.d. and 250 µm i.d. × 365 µm o.d. polyimide-coated fused silica (Polymicro Technologies, Phoenix, AZ).
2. Model P-2000 Laser Puller (Sutter Instrument Co. Novato, CA).
3. Stainless steel pressurization device (Brechbuehler, Inc., Houston, TX, or MTA for blueprints available by request from John Yates, Scripps Research Institute, La Jolla, CA).
4. Five-micrometer C_{18} Aqua Reversed Phase Packing Material (Phenomenex, Torrance, CA).
5. 5-µm Partisphere SCX packing material (Whatman, Florham Park, NJ).
6. M-520 Inline Micro Filter Assembly and F-185 Microtight 0.0155 × 0.025 Sleeves (UpChurch Scientific, Oak Harbor, WA).

2.3. Multidimensional Chromatography and Tandem Mass Spectrometry

1. Buffer A: 5% acetonitrile, 0.1% formic acid, made with HPLC-grade water.
2. Buffer B: 80% acetonitrile, 0.1% formic acid, made with HPLC-grade water.
3. Buffer C: 500 m*M* ammonium acetate, 5% acetonitrile, 0.1% formic acid, made with HPLC-grade water.
4. 50 μm i.d. × 365 × μm o.d. polyimide-coated fused silica (Polymicro Technologies, Phoenix, AZ).
5. P-775 MicroTee Assemblies and F-185 Microtight 0.0155 × 0.025 Sleeves (UpChurch Scientific, Oak Harbor, WA).
6. Gold wire (Scientific Instrument Services, Inc., Ringoes, NJ).
7. Agilent 1100 series G1379A degasser, G1311A quaternary pump, G1329A autosampler, G1330B autosampler thermostat, and G1323B controller (Agilent Technologies, Palo Alto, CA).
8. LCQ DECA-XP^{plus} tandem mass spectrometer (Thermo Electron, San Jose, CA).
9. Nano electrospray stage (MTA for blueprints available by request from John Yates, Scripps Research Institute, La Jolla, CA). Other options include the Thermo Electron Nanospray II ion source or PicoView Sources from New Objective (Woburn, MA).

3. Methods

3.1. Sample Preparation

A wide variety of samples have been analyzed via MudPIT, including protein complexes ***(13)***, organelles ***(14)***, whole proteomes ***(7,8)***, and biofluids ***(15)***. The isolation of protein mixtures from any of these samples is beyond the scope of this article. It is important to have a good idea what the total amount of protein in a sample is, however, in order to add the proper amount of proteases for digestion. In general, all of these samples from disparate sources will be analyzed in the fashion outlined in this chapter, with the only variation being that if a large amount of proteins is in the starting sample, it will be split into 200- to 300-μg aliquots and each fraction will be TCA precipitated independently and then recombined after resuspension for protease digestion. The standard approach for the detection and identification of proteins is the endoproteinase LysC/trypsin digestion protocol. For the identification of posttranslational modifications, high sequence coverage of individual proteins is needed. In this case, samples will be split into at least three fractions and independently digested with endoproteinase LysC/trypsin, elastase, and subtilisin in a similar fashion to the protocol published by MacCoss et al. ***(16)***. The high-pH proteinase K digestion protocol can also be used to generate high sequence-coverage data from complex protein mixtures for post-translational modification analysis ***(14)***.

3.1.1. TCA Precipitation of Proteins From Solutions (see **Notes 1–4**)

1. Bring the final sample solution of a cellular extract or protein complex, for example, to 400 μL with 100 m*M* Tris-HCl (pH 8.5). If a larger volume of final sample solution is available, split the solution into multiple 200-μL aliquots followed by the addition of 100 m*M* Tris-HCl (pH 8.5) to 400 μL. At the resuspension step (**Subheading 3.1.2., step 1**), the resolubilized precipitated proteins may be recombined.
2. Add 100 μL TCA (100%) to cold sample; mix well to give a final TCA concentration of 20%. Reaction should be carried out on ice and the sample left overnight at 4°C.
3. Spin at 20,800*g* for 30 min at 4°C; aspirate the supernatant with gel loading tip, leaving 5 μL in the tube so as not to disturb the pellet.
4. Wash with 2 × 500 μL of cold acetone. After each wash, spin for 10 min at 20,800*g*.
5. Dry using a speed vac for 5 min.

3.1.2. Protein Denaturation, Reduction, and Alkylation

1. Add 100 m*M* Tris-HCl (pH 8.5), 8 *M* urea to TCA precipitated proteins; vortex.
2. Bring solution to 5 m*M* TCEP with the 0.1 *M* TCEP solution (1 *M* stock diluted 1/10); incubate at room temperature for 30 min.
3. Bring solution to 10 m*M* IAM with 0.5 *M* stock; incubate at room temperature for 30 min in dark.

3.1.3. Endoproteinase LysC/Trypsin Digestion

1. Add endoproteinase LysC at 1 μg/μL (1:100) to the denatured, reduced, and carboxymethylated proteins; incubate at 37°C for at least 6 h.
2. Dilute to 2 *M* urea with 100 m*M* Tris-HCl (pH 8.5).
3. Add $CaCl_2$ to 2 m*M* (stock at 500 m*M*).
4. Add trypsin at 0.1 μg/μL (1:100); incubate at 37°C overnight while shaking.
5. On the next day, add 90% formic acid to 5%.
6. Store sample at –80°C.

3.1.4. Elastase Digestion

1. Dilute the denatured, reduced, and carboxymethylated proteins to 2 *M* urea with 100 m*M* Tris-HCl (pH 8.5).
2. Add elastase to an enzyme-to-substrate ratio of 1:50 (w/w); incubate at 37°C for 6 h while shaking.
3. Add 90% formic acid to 5%.
4. Store sample at –80°C.

3.1.5. Subtilisin A Digestion

1. Dilute the denatured, reduced, and carboxymethylated proteins to 4 *M* urea with 100 m*M* Tris-HCl (pH 8.5).
2. Add subtilisin A to an enzyme-to-substrate ratio of 1:50 (w/w); incubate at 37°C for 2–3 h while shaking.

3. Add 90% formic acid to 5%.
4. Store sample at –80°C.

3.1.6. High-pH Proteinase K Digestion

1. Resuspend TCA-precipitated proteins in 100 m*M* sodium carbonate (pH 11.5).
2. Add solid urea to 8 *M* urea; vortex.
2. Bring solution to 5 m*M* TCEP with the 0.1 *M* TCEP solution (1 *M* stock diluted 1/10); incubate at room temperature for 30 min.
3. Bring solution to 10 m*M* IAM with 0.5 *M* stock; incubate at room temperature for 30 min in dark
4. Add proteinase K at 0.25 µg/µL, at an enzyme-to-substrate ratio of 1:100 (w/w); incubate at 37°C for 4 h while shaking.
5. Add 90% formic acid to 5%.
6. Store sample at –80°C.

3.2. Microcapillary Column Construction and Sample Loading

Currently, all samples are desalted on-line using columns similar to the three-phase microcapillary columns described in McDonald et al. ***(5)***. These columns contain reversed-phase material, followed by strong cation exchange material, followed by reversed-phase material. In an abbreviated fashion, they are RP/SCX/RP columns. By using these columns, one does not need to carry out additional sample cleanup and buffer exchange prior to loading. For sample quantities of 400 µg or less, the triple-phase fused-silica microcapillary column is used, and for samples containing more than 400 µg, the split triple-phase fused-silica microcapillary column is used. In addition, the split triple-phase column is used for samples that originally contained detergents, allowing for more extensive washing after sample loading.

3.2.1. Pulling Columns

1. Make a window in the center of approx 50 cm of 100 µm × 365 µm fused-silica capillary by holding it over an alcohol flame until the polyimide coating has been charred. The charred material is removed by gently wiping the capillary with a tissue soaked in methanol.
2. To pull a needle, place the capillary into the P-2000 laser puller. Position the exposed window of the capillary in the mirrored chamber of the puller. Arms on each side of the mirror have grooves and small vises, which properly align the fused silica and hold it in place. Our four-step parameter setup for pulling approx 3-µm tips from a 100 µm i.d. × 365 µm o.d. capillary is as follows with all other values set to zero:

   Heat = 290, Velocity = 40, and Delay = 200
   Heat = 280, Velocity = 30, and Delay = 200
   Heat = 270, Velocity = 25, and Delay = 200
   Heat = 260, Velocity = 20, and Delay = 200

3.2.2. Triple-Phase Fused-Silica Microcapillary Column

1. Pull tip with laser puller as described under **Subheading 3.2.1.**
2. Place approx 20 mg of Aqua RP packing material into a 1.7-mL microfuge tube, add 1 mL of MeOH, and place the tube into a stainless steel pressurization vessel. Secure the pressurization vessel lid by tightening the bolts.
3. The lid has a Swagelok® fitting containing a 0.4-mm Teflon ferrule. Feed the fused-silica capillary (pulled end up) down through the ferrule until the end of the capillary reaches the bottom of the microfuge tube. Tighten the ferrule to secure the capillary.
4. Apply pressure to the pressurization vessel by first setting the regulator on the gas cylinder to approx 400–800 psi, then opening a valve on the pressurization vessel to pressurize it. The packing material will begin filling the pulled needle capillary. If it does not flow right away, gently open the tip using a capillary scriber. Pack the capillary with 9 cm of RP packing material (*see* **Note 5**).
5. Slowly release the pressure from the pressurization vessel so as to not cause the packed RP material to unpack. Open the stainless steel pressurization vessel and remove the microfuge tube containing the RP material in MeOH.
6. Place approx 20 mg of Whatman Paritshpere SCX packing material into a 1.7-mL microfuge tube, add approx 1 mL of MeOH, and place the tube into the pressurization vessel. Secure the pressurization vessel as described.
7. Apply pressure as described in **Subheading 3.2.2., step 4**. Pack the capillary with 3 cm of the SCX material and then slowly release the pressure as described.
8. Place the tube with Aqua RP particles in methanol back into the pressurization vessel and add 2–3 cm of RP material after the SCX material.
9. Wash with methanol for at least 10 min.
10. Equilibrate with buffer A (5% ACN, 0.1% formic acid) for at least 30 min.
11. To get rid of any particulates (which could clog the microcapillary column), spin down the sample to be loaded for 30 min at 20,800*g*, and transfer the supernatant to a new 1.7-mL microfuge tube.
12. Sample can then be loaded by placing the 1.7-mL microfuge tube into the pressurization vessel.
13. After sample is loaded, 500 μL of buffer A is added to the microfuge tube that contained the sample, and the loaded column is washed until installed onto the mass spectrometer (at least 1 h).

3.2.3. Split Triple-Phase Fused-Silica Microcapillary Column

1. The first step in this process is to prepare a double-phase column out of 250-μm fused silica microcapillary. Place approx 2 × 15 cm of 250 μm i.d. × 365 μm o.d. fused silica capillaries on both sides of an M-520 Inline Micro Filter Assembly.
2. As described under **Subheading 3.2.2.**, first pack the column with 3–4 cm of Whatman Partisphere SCX packing material by leading the 250-μm capillary connected to the fritted side of the inline micro-filter assembly into the pressurization vessel. The capillary connected to the other end of the inline assembly serves as a waste line flowing into an empty microfuge tube.

3. Then pack with 2–3 cm of Phenomenex Aqua RP packing material.
4. Wash with methanol for at least 10 min.
5. Equilibrate in buffer A for at least 30 min.
6. Sample can then be loaded by placing the 1.7-mL microfuge tube into the pressurization vessel.
7. After sample is loaded, 1.5 mL of buffer A is added to the microfuge tube that contained the sample, and the loaded column is washed until it is installed onto the mass spectrometer. Because the split three-phase columns are used for larger sample amounts and for samples that contained detergent, longer washing is necessary. Ideally, all 1.5 mL of buffer A is used for washing.
8. Next, a single-phase RP column, to be added to the other end of the filtered union assembly, is prepared by first pulling a tip using 100 μm i.d. × 365 μm o.d. fused silica with the laser puller as described under **Subheading 3.2.1.** and packing 9 to 10 cm of RP material.
9. This single-phase RP column is equilibrated with buffer A (5% ACN, 0.1% formic acid) for at least 30 min.
10. When both parts of the split column have been washed, the single-phase 100-μm RP column can be connected to the loaded 250-μm double-phase capillary using the filtered union.

3.3. Multidimensional Chromatography and Tandem Mass Spectrometry

The MudPIT system we use is a combination of the Agilent1100 quaternary pump stack with thermostatted autosampler, LCQ DECA-XP^{plus} tandem mass spectrometer, and a nanoelectrospray stage that interfaces the two systems. Furthermore, the hand-made microcapillary columns described above are single use, and the fused silica portion of each column is discarded after every analysis (*see* **Note 6**).

3.3.1. Setup of the Nanoelectrospray Stage

1. A detailed schematic of the nanoelectrospray stage is shown in **Fig. 1**. The triphasic microcapillary column or split three-phase column should be attached as shown to the P-775 MicroTee Assemblies.
2. One connection point of the cross contains the transfer line from the HPLC pump. This consists of a piece of 100 μm i.d. × 365 μm o.d. polyimide-coated fused silica.
3. A second connection point contains a length of 50 μm i.d. × 365 μm o.d. fused silica capillary that is used as a split/waste line. This split line allows a majority of the flow to exit through the split; therefore, very low flow rates can be achieved through the packed capillary micro-column. The size and length of this section of capillary depend on the flow rate from the pump and the length of the micro-column. A good starting point is to use a 12-in section of 50 μm i.d. × 365 μm o.d. polyimide-coated fused silica for the split line.
4. Another connection contains a section of gold wire. This is to allow the solvent entering the needle to be energized to 2400 V, thus allowing electrospray ionization to occur.

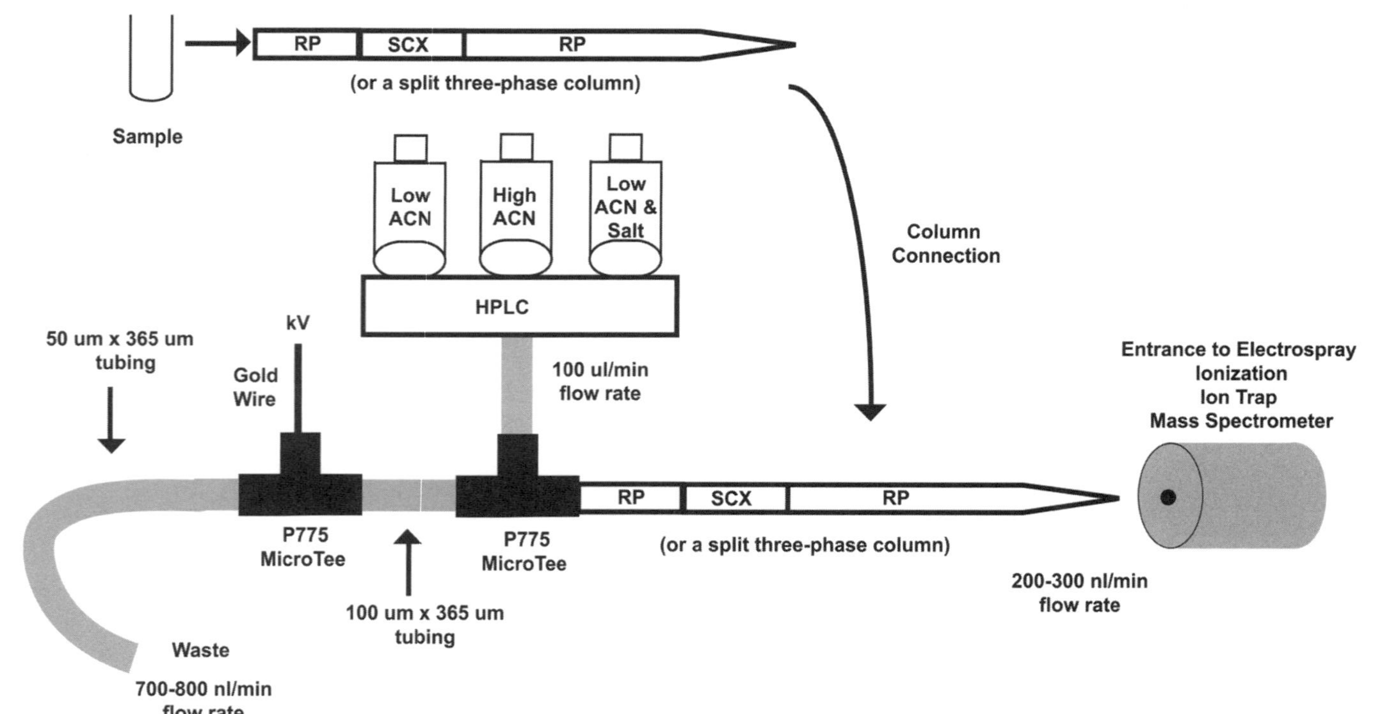

Fig. 1. Three-phase multidimensional protein identification technology (MudPIT) column setup. In the MudPIT system, a triphasic microcapillary column with both strong cation exchange (SCX) and reversed-phase (RP) packing materials is prepared and loaded off-line. Upon insertion of the system, a fully automated analysis can be run with salt bumps moving fractions from the RP to the SCX by a first step reversed-phase gradient and to the to the RP using a series of salt bumps detailed in **Tables 1–3**. Upon the application of RP gradients, peptides elute into the mass spectrometer. In this approach, a volatile salt must be used, like ammonium acetate or ammonium formate.

5. Place the packed, loaded, and washed column into a MicroTee on a stage, which in this case is designed for the ThermoFinnigan DECA-XP^{plus} series mass spectrometer. This stage performs a threefold purpose: to support the MicroTee and hold it in place along with the connections, to electrically insulate the MicroTee from contact with its surroundings when it is held at high voltage potential, and to allow for fine position adjustments of the micro-column with respect to the entrance of the mass spectrometer (heated capillary) by using an XYZ manipulator.
6. Measure the flow from the tip of the capillary micro-column, using graduated glass capillaries. To do this, set the flow rate of the Agilent1100 to 0.1 mL/min from the controller. The target flow rate at the tip should be approx 200–300 nL/min and a back pressure on the Agilent1100 of between 30 and 50 bars. If the flow rate is too high, cut off a portion of the split line capillary. This will cause more of the flow to exit out of the split and cause less flow through the micro-column. If the flow is too low, a longer piece of 50-μm capillary or a section with a smaller inner diameter can be used to force more flow through the micro-column. Measuring the flow rate and adjusting the split line may have to be repeated a number of times until the target flow rate is reached (*see* **Notes 7** and **8**).
7. Prior to initiating a run, position the micro-column using the XYZ manipulator so that the needle tip is within 5 mm from the orifice of the mass spectrometer's heated capillary.

3.3.2. Instrument Method Design Description (see ***Note 9***)

Data-dependent acquisition of tandem mass spectra during the HPLC gradient is also programmed through the LCQ Xcalibur™ software. Here we provide guidance for setting the parameters for data-dependent acquisition using the DECA-XP^{plus} series mass spectrometer. The following settings are for a typical data-dependent MS/MS acquisition analysis. The method consists of a continual cycle beginning with one scan of MS (scan one), which records all of the *m/z* values of the ions present at that moment in the gradient, followed by three rounds of MS/MS. Full MS spectra are recorded on the peptides over a 400 to 1600 *m/z* range. Dynamic exclusion is activated to improve the protein identification capacity during the analysis.

1. In the main Xcalibur software page, select "Instrumental Setup."
2. In the next window, select the button labeled "Data Dependent MS/MS."
3. The general settings for any given method is as follows: "segments" is set to 1; "start delay" is set to 0; "duration" is 117 min; "number of scan events" is 4; "scan event details" are first, MS, second, MS/MS of most intense ion, third, MS/MS of second most intense ion, fourth, MS/MS of third most intense ion.
4. For scan event one, highlight the bar showing "Scan Event 1." Below this, check "Normal Mass Range"; "Scan Mode" is MS, and "Scan Type" is full. Set "m/z range" to 400–1400, "Polarity" to positive, and "Data Type" to centroid.
5. The "Tune Method" box specifies the path for a file containing the parameters for the electrostatic lenses and ion trap. These parameters are established through the

"LCQ Tune" as described in the "ThermoFinnigan LCQ Getting Started" manual. The capillary temperature is 200°C and the electrospray voltage is 2.4 kV.

6. The data-dependent settings are as follows, with only these options engaged: "default charge state" is 2, "default isolation width" is 3, "normalized collision energy" is 35, "activation time" is 30, "minimal signal required" is 100,000, "minimal MSn signal required" is 5000, "exclusion mass width low" is 0.80, and "exclusion mass width high" is 2.20. Dynamic exclusion is enabled, with the following settings: a repeat count of 2, a repeat duration of 0.5, an exclusion list of 50, an exclusion duration of 5.00 min, an exclusion mass width low of 0.80, and an exclusion mass width high of 2.20.
7. In the "Timed Events" window of the instrument setup, the following sequence should be inserted:

   Time (min) = 0.00, Settings = Contact 1, Value = Open
   Time (min) = 0.05, Settings = Contact 1, Value = Closed
   Time (min) = 0.10, Settings = Contact 1, Value = Open

3.3.3. Gradient Profiles for Complex Mixture Analysis

We use two general gradient profiles. One is for analyzing protein complexes and one is for analyzing organelles or other cellular fractions. The method for analyzing protein complexes takes approx 10 h, and the second method takes approx 20 h. Because we use three-phase columns, the first step in any run is a reversed-phase gradient to move any bound peptide from the first RP to the SCX material inside the column. Then, successive salt bumps are run to move small amounts of peptides from the SCX onto the last RP, followed by a slow reversed-phase gradient to resolve peptides within the last RP before they are eluted off into the mass spectrometer. In the "Gradient Program" window of the instrument setup, the following sequences are steps that can be used or adjusted by a researcher. An example base peak chromatogram is shown in **Fig. 2** as an example of the profile that should be seen from one step of a successful sample.

Table 1
Gradient Profile of First Step

Time (min)	Flow Rate (mL/min)	% Buffer A	% Buffer B
0	0.1	100	0
16	0.1	60	40
17	0.1	0	100
20	0.1	0	100

This is the general gradient profile that we use for all additional steps where the % of buffer C increases with each step:

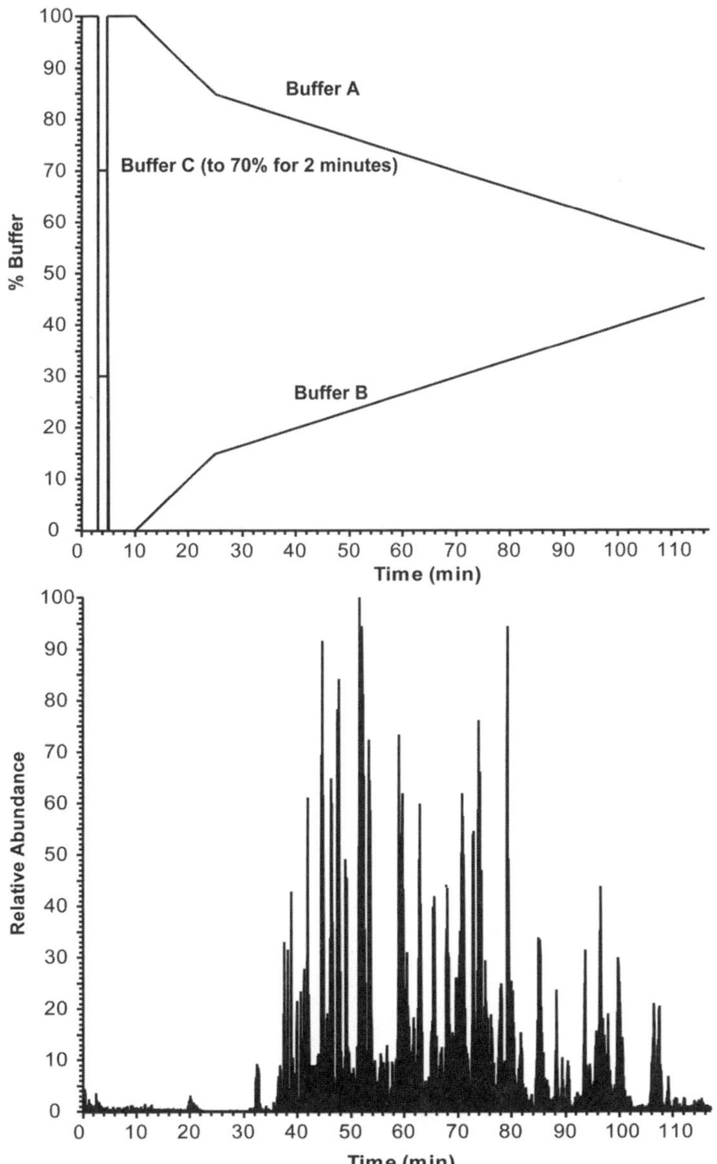

Fig. 2. Chromatographic profile of the fifth step of a multidimensional protein identification technology (MudPIT) analysis of the HeLa cell mediator component CRSP70. As described in Sato et al. *(13)*, the HeLa cell mediator component CRSP70 was FLAG tagged and affinity purified from HeLa cell extracts and analyzed via the six-step MudPIT chromatographic elution profile shown in **Table 4**. **(A)** The % buffer composition of buffers A, B, and C over the 117-min step 5, which contains a 70% buffer C salt bump. **(B)** The base peak chromatogram (*m/z* 400–1400) of this step for the CRSP70 analysis, which is a general representation of the peptides eluting from the biphasic MudPIT column into a tandem mass spectrometer.

Table 2
Gradient Profile of X(2–80)% Buffer C Step

Time (min)	Flow Rate (mL/min)	% Buffer A	% Buffer B	% Buffer C
0	0.1	100	0	0
3	0.1	100	0	0
3.1	0.1	98	0	X
5	0.1	98	0	X
5.1	0.1	100	0	0
10	0.1	100	0	0
10.1	0.1	100	0	0
25	0.1	85	15	0
117	0.1	55	45	0

This is the last step for any given MudPIT analysis, and is designed to try to remove everything off the column:

Table 3
Gradient Profile of 100% Buffer C Step

Time (min)	Flow Rate (mL/min)	% Buffer A	% Buffer B	%Buffer C
0	0.1	100	0	0
2	0.1	100	0	0
2.1	0.1	0	0	100
22	0.1	0	0	100
22.1	0.1	100	0	0
27	0.1	100	0	0
37	0.1	80	20	0
85	0.1	30	70	0
90	0.1	0	100	0
90.1	0.1	0	100	0
95	0.1	0	100	0
95.1	0.1	100	0	0
97	0.1	100	0	0

In the Xcalibur software main page, check the "sample list" button. In the "Sequence Setup" window, fill in the "Filename," "Path," and "Inst Method" lines with the appropriate methods indicated previously. Save the sequence to the same directory of the paths for the *.RAW files. In the table below are the sequences we use for the analysis of a protein complex (6-step MudPIT) and for more complex mixtures like an organelle (12-step MudPIT):

Table 4
Multidimensional Protein Identification Technology (MudPIT) Analyses Methods

Method	6 step MudPIT	12 Step MudPIT
20 min reverse phase	Yes	Yes
2% Buffer C		Yes
6% Buffer C		Yes
10% Buffer C		Yes
15% Buffer C	Yes	Yes
20% Buffer C		Yes
30% Buffer C	Yes	Yes
40% Buffer C		Yes
50% Buffer C	Yes	Yes
60% Buffer C		Yes
70% Buffer C	Yes	
80% Buffer C		Yes
100% Buffer C	Yes	Yes

5. Once everything is set up and the needle is positioned correctly in front of the heated capillary on the mass spectrometer, hit "Actions" and "Run Sequence." A box will come up that should have the "Agilent1100 Quat," "Agilent 1100 Thermostatted AS," and "LCQ DECAXP MS" listed, and "Yes" should be typed under the "Start Instrument" box of the "Agilent 1100 Thermostatted AS." The "start when ready" box should be checked, as well as "After Sequence Set System to –Standby." Both the "pre-acquisition" and "post-acquisition" boxes under "run synchronously" should contain check marks. Hit "OK" and the run will begin. A message will likely come up saying that there are devices that need to be turned on; hit "OK."
6. Upon the completion of a run, the *.RAW files that have been accumulated by the mass spectrometer need to be converted to *.DAT files for SEQUEST analysis. To do this, go to the "Home Page" and hit the "Tools" and "File Converter" buttons. Follow the instructions to select the jobs that need to be converted to *.DAT files, and hit "OK."

3.4. Data Analysis

For complex peptide mixture analysis via MudPIT, database searching is an important component and requires a computer cluster to analyze tandem mass spectra. SEQUEST *(6)* clusters are available from IBM and ThermoElectron (San Jose, CA), and Mascot *(17)* clusters are available from Matrix Science (Boston, MA). In addition, researchers who have the necessary expertise can design their own cluster and purchase licenses for these software platforms to install on their cluster. Lastly, an open-source tandem mass spectral searching algorithm named X!Tandem ***(18,19)*** is emerging as an alternative to SEQUEST ***(6)*** and Mascot ***(17)***. There is growing evidence in the literature that single-peptide hits to proteins from complex mixtures must be considered with great care, but proteins that are identified by at least two good-quality tandem mass spectra are likely to be real identifications ***(20,21)***. The manual assessment criteria listed below offer suggestions on how to evaluate spectrum/peptide matches.

3.4.1. SEQUEST Analysis

1. The software algorithm 2to3 ***(22)*** is used to determine charge state and to delete spectra of poor quality.
2. SEQUEST ***(6)*** is used to match MS/MS spectra to peptides in a database containing protein sequences (typically *.fasta files downloaded from the National Center for Biotechnology Information and updated frequently), complemented with sequences for common contaminants, including human keratin variants and immunoglobulins.
3. The validity of peptide/spectrum matches is assessed using the SEQUEST-defined parameters, cross-correlation score (XCorr), and normalized difference in cross-correlation scores (DeltCn). Spectra/peptide matches are retained only if they have a DeltCn of at least 0.08 and minimum XCorr of 1.8 for singly, 2.5 for doubly, and 3.5 for triply charged spectra. In addition, the peptides have to be at least seven amino acids long.
4. The program DTASelect ***(23)*** is used to select and sort peptide/spectrum matches passing this criteria set. Peptide hits from multiple runs are compared using CONTRAST ***(23)***.

3.4.2. Manual Assessment Criteria

1. The MS/MS spectrum must be of good quality, with fragment ions clearly above baseline noise. Sometimes a spectrum will consist purely of "sticks." In general, these are poor-quality spectra. There should be some noise that allows one to define what the baseline noise is.
2. There must be some continuity to the *b* and *y* ion series. Of ions that are matched by SEQUEST to the original tandem mass spectrum, the matched ions should correspond to signal rather than noise. Furthermore, better matches contain strings of ions rather than isolated ions here and there throughout the sequence.
3. The intensity of a *b* or *y* ion resulting from a proline should be far more intense than the other ion in the spectrum. If there are multiple prolines present in a peptide, one may see internal fragment ions corresponding to the mass of the portion of the peptide between the two prolines.
4. When evaluating peptides generated by nontryptic digestions, be aware that because basic residues can be located anywhere within the peptides, doubly charged precursor ions can lead to doubly charged fragments, which will appear as unmatched intense peaks in most spectrum displays ***(24)***.

4. Notes

1. For samples that likely contain large amounts of DNA, the addition of 0.1 U of Benzonase to the sample prior to TCA precipitation and incubation at 37°C for 30 min is critical. This will decrease the viscosity of the mixture and prevent column clogging during sample loading, washing, or analysis.
2. Using the TCA precipitation protocol as the starting point, the only chemical that cannot be dealt with is glycerol. The TCA precipitation protocol has been useful in detergent removal. When the presence of glycerol cannot be avoided, loading such samples on split columns with extensive washes may prevent column clogging.
3. With the TCA precipitation protocol and protein complexes, it is common to not see a pellet. For this reason, a small amount of acetone should be left at the bottom of the tube during washing to be sure to not aspirate and discard the pellet.
4. When handling samples, gloves must be worn at all times in order to prevent the addition of large amounts of human keratin and immunoglobulins to a sample.
5. Placing a black three-ring binder about 6 in. behind the loading column will provide the contrast necessary to see the status of column packing.
6. The microassemblies used for the split three-phase columns can be reused by first sonicating in 50% methanol:50% water and allowing them to air dry. However, we discard the filters (M-120X filter end fitting from UpChurch Scientific, Oak Harbor, WA) and use a new one for each sample.
7. If the Agilent1100 reaches the high-pressure limit (set at 100 bars), the most likely place for clogging is the small piece of 100 μm i.d. × 365 μm o.d. fused-silica tubing that connects the two MicroTees (**Fig. 1**). If changing the capillary does not alleviate the pressure, removing the MicroTees and sonicating them in 50% methanol:50% water and allowing them to air dry may solve the problem.
8. For columns that clog during loading of the sample, washing of the sample, or testing of the flow rate prior to analysis, heating the length of the column with a Varitemp Heat Gun (model VT-750C, Master Appliance Corp, Racine, WI) usually makes it flow again. Once a column starts flowing again, extensive additional washing with buffer A is recommended to prevent subsequent clogging.
9. Especially with whole cell extracts and their analysis by MudPIT, it is critical to clean and tune the mass spectrometer according to the manufacturer's instructions at least every 10 d.

References

1. Evans, C. R. and Jorgenson, J. W. (2004) Multidimensional LC-LC and LC-CE for high-resolution separations of biological molecules. *Anal. Bioanal. Chem.* **378,** 1952–1961.
2. Link, A. J., Eng, J., Schieltz, D. M., et al. (1999) Direct analysis of protein complexes using mass spectrometry. *Nat. Biotechnol.* **17,** 676–682.
3. Washburn, M. P., Wolters, D., and Yates, J. R., 3rd. (2001) Large-scale analysis of the yeast proteome by multidimensional protein identification technology. *Nat. Biotechnol.* **19,** 242–247.

4. Wolters, D. A., Washburn, M. P., and Yates, J. R., 3rd. (2001) An automated multidimensional protein identification technology for shotgun proteomics. *Anal. Chem.* **73,** 5683–5690.
5. McDonald, W. H., Ohi, R., Miyamoto, D. T., Mitchison, T. J., and Yates, J. R. (2002) Comparison of three directly coupled HPLC MS/MS strategies for identification of proteins from complex mixtures: single-dimension LCMS/MS, 2-phase MudPIT, and 3-phase MudPIT. *Int. J. Mass Spectrom.* **219,** 245–251.
6. Eng, J., McCormack, A. L., and Yates, J. R., III. (1994) An approach to correlate tandem mass spectral data of peptides with amino acid sequences in a protein database. *J. Amer. Mass Spectrom.* **5,** 976–989.
7. Florens, L., Washburn, M. P., Raine, J. D., et al. (2002) A proteomic view of the *Plasmodium falciparum* life cycle. *Nature* **419,** 520–526.
8. Koller, A., Washburn, M. P., Lange, B. M., et al. (2002) Proteomic survey of metabolic pathways in rice. *Proc. Natl. Acad. Sci. USA* **99,** 11,969–11,974.
9. Pan, Y., Kislinger, T., Gramolini, A. O., et al. (2004) Identification of biochemical adaptations in hyper- or hypocontractile hearts from phospholamban mutant mice by expression proteomics. *Proc. Natl. Acad. Sci. USA* **101,** 2241–2246.
10. Washburn, M. P., Koller, A., Oshiro, G., et al. (2003) Protein pathway and complex clustering of correlated mRNA and protein expression analyses in Saccharomyces cerevisiae. *Proc. Natl. Acad. Sci. USA* **100,** 3107–3112.
11. Wu, C. C., MacCoss, M. J., Howell, K. E., and Yates, J. R., 3rd. (2003) A method for the comprehensive proteomic analysis of membrane proteins. *Nat. Biotechnol.* **21,** 532–538.
12. Graumann, J., Dunipace, L. A., Seol, J. H., et al. (2004) Applicability of tandem affinity purification MudPIT to pathway proteomics in yeast. *Mol. Cell. Proteomics* **3,** 226–237.
13. Sato, S., Tomomori-Sato, C., Parmely, T. J., et al. (2004) A set of consensus mammalian mediator subunits identified by multidimensional protein identification technology. *Mol. Cell* **14,** 685–691.
14. Wu, C. C., MacCoss, M. J., Mardones, G., et al. (2004) Organellar proteomics reveals golgi arginine dimethylation. *Mol. Biol. Cell* **15,** 2907–2919.
15. Fujii, K., Nakano, T., Kawamura, T., et al. (2004) Multidimensional protein profiling technology and its application to human plasma proteome. *J. Proteome Res.* **3,** 712–718.
16. MacCoss, M. J., McDonald, W. H., Saraf, A., et al. (2002) Shotgun identification of protein modifications from protein complexes and lens tissue. *Proc. Natl. Acad. Sci. USA* **99,** 7900–7905.
17. Perkins, D. N., Pappin, D. J., Creasy, D. M., and Cottrell, J. S. (1999) Probability-based protein identification by searching sequence databases using mass spectrometry data. *Electrophoresis* **20,** 3551–3567.
18. Craig, R., Cortens, J. P., and Beavis, R. C. (2004) Open source system for analyzing, validating, and storing protein identification data. *J. Proteome Res.* **3,** 1234–1242.
19. Craig, R. and Beavis, R. C. (2004) TANDEM: matching proteins with tandem mass spectra. *Bioinformatics* **20,** 1466–1467.

20. Carr, S. A., Aebersold, R., Baldwin, M., Burlingame, A., Clauser, K., and Nesvizhskii, A. (2004) The need for guidelines in publication of peptide and protein identification data: working group on publication guidelines for peptide and protein identification data. *Mol. Cell. Proteomics* **3,** 531–533.
21. Venable, J. D. and Yates, J. R., 3rd. (2004) Impact of ion trap tandem mass spectra variability on the identification of peptides. *Anal. Chem.* **76,** 2928–2937.
22. Sadygov, R. G., Eng, J., Durr, E., et al. (2002) Code developments to improve the efficiency of automated MS/MS spectra interpretation. *J. Proteome Res.* **1,** 211–215.
23. Tabb, D. L., McDonald, W. H., and Yates, J. R., 3rd. (2002) DTASelect and Contrast: tools for assembling and comparing protein identifications from shotgun proteomics. *J. Proteome Res.* **1,** 21–26.
24. Tabb, D. L., Huang, Y., Wysocki, V. H., and Yates, J. R., 3rd. (2004) Influence of basic residue content on fragment ion peak intensities in low-energy collision-induced dissociation spectra of peptides. *Anal. Chem.* **76,** 1243–1248.

12

Isolation of Glycoproteins and Identification of Their *N*-Linked Glycosylation Sites

Hui Zhang and Ruedi Aebersold

Summary

Protein glycosylation has long been recognized as a very common posttranslational modification. Protein glycosylation is prevalent in proteins destined for extracellular environments. These include proteins localized on the extracellular surface and those secreted to body fluids. In search of a method that has the potential to identify and quantify most proteins found in body fluids or the cell surface, we have recently developed a novel method for solid-phase extraction of formerly *N*-linked glycosylated peptides from glycoproteins. It has been shown that proteins secreted to body fluids or localized on the cell surface can be specifically enriched by this method. The technique is based on the conjugation of glycoproteins to a solid support using hydrazide chemistry, removal of nonglycosylated peptides by trypsin digestion, stable isotope labeling of glycopeptides, and the specific release of formerly *N*-linked glycosylated peptides via peptide-*N*-glycosidase. The recovered formerly *N*-linked glycopeptides are then identified and quantified by tandem mass spectrometry.

Key Words: Glycoprotein; glycopeptide; *N*-linked glycosylation; proteomics; glycoproteomics; mass spectrometry; protein identification; membrane protein; secreted protein; plasma; serum; body fluid; cell surface proteins; immunotherapy target; biomarker.

1. Introduction

Protein glycosylation is the most common posttranslational modification of proteins that are destined for extracellular environments *(1)*, and it has been shown that glycosylation plays a critical role in protein interactions, folding, protein stabilization, subcellular localization, and protein function *(2)*. Typically, carbohydrates are linked to the side chains of serine or threonine residues (*O*-linked glycosylation) or to asparagine residues (*N*-linked glycosylation) *(2)*. *N*-linked glycosylation sites generally fall into the N-X-S/T sequence motif, in which X denotes any amino acid except proline *(3)*. There is no consensus motif

From: *Methods in Molecular Biology, Vol. 328: New and Emerging Proteomic Techniques*
Edited by: D. Nedelkov and R. W. Nelson © Humana Press Inc., Totowa, NJ

that has been identified to date for *O*-linked glycosylation. Protein glycosylation, and in particular *N*-linked glycosylation, is prevalent in extracellular surface proteins, secreted proteins, and proteins contained in body fluids (blood serum, cerebrospinal fluid, urine, breast milk, saliva, lung lavage fluid, pancreatic juice, and so on). These also happen to be the specimens in the human body that are most easily accessible for diagnostic or therapeutic purposes. It is therefore no surprise that many clinical biomarkers and therapeutic targets are glycoproteins. These include Her2/neu in breast cancer, human chorionic gonadotropin and α-fetoprotein in germ cell tumors, PSA in prostate cancer, CA125 in ovarian cancer, and CEA for colon, breast, pancreatic, and lung cancer ***(4)***. The Her2/neu receptor is also the target for a successful immunotherapy of breast cancer using the humanized monoclonal antibody Herceptin ***(5)***. It is believed that the majority of serum-specific proteins, cell-surface proteins, and secreted proteins are in fact glycosylated ***(6)***. Therefore, the specific isolation of glycopeptides from serum, tissues, and cells focuses on an information-rich protein population that is likely to contain useful immunotherapy targets on the cell surface or disease biomarkers in body fluids ***(7)***.

We have recently developed a novel method for solid-phase extraction of formerly *N*-linked glycosylated peptides from glycoproteins (SPEG) and shown that it can specifically enrich proteins secreted to body fluids or localized on the cell surface ***(8–10)***. The method we describe here in detail is a proteome-wide approach to identify glycoproteins and their *N*-linked glycosylation sites. The principle of this method is that glycoproteins can be conjugated to a solid support via their carbohydrate chains, using hydrazide chemistry. The *cis*-diol groups in oligosaccharide chains of glycoproteins can be oxidized by periodate oxidation to produce di-aldehydes, which are reactive to hydrazide on a solid support to form covalent hydrazone bonds. After removal of the nonglycosylated peptides by protease digestion and washing, the immobilized glycopeptides are released from the solid support by *N*-linked glycosylation-specific enzymatic cleavage. The isolated peptides are analyzed by mass spectrometry. The glycopeptides can also optionally be isotopically labeled at their amino termini to allow the quantities of glycopeptides isolated from different biological samples to be compared.

The procedure of SPEG is schematically illustrated in **Fig. 1**. The method includes the following steps: (1) glycoprotein oxidation; (2) coupling of glycoproteins to a solid support; (3) proteolysis—the immobilized glycoproteins are proteolyzed on the solid support and the nonglycosylated peptides are removed by washing, whereas the glycosylated peptides remain on the solid support; (4) isotopic labeling (optional)—the amino groups of the immobilized glycopeptides are labeled with the light (d0, contains no deuteriums) or heavy (d4, contains four deuteriums) forms of succinic anhydride after conversion of the ε-amino groups of lysine to homoarginine; (5) release—formerly *N*-linked gly-

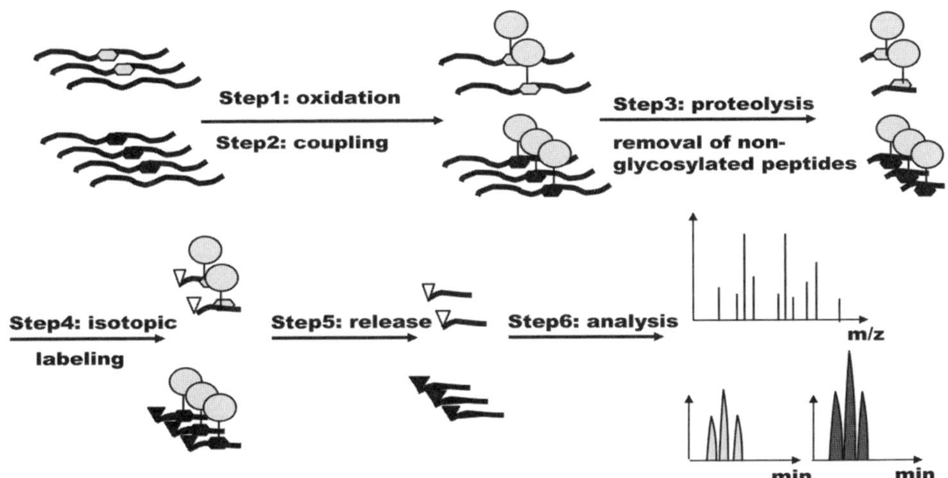

Fig. 1. Schematic diagram of an exemplary method of isolation, identification, and quantification of glycopeptides.

copeptides are released from the oligosaccharides and solid-phase support by peptide-*N*-glycosidase F (PNGase F) treatment; (6) analysis—the formerly *N*-linked glycopeptides and corresponding proteins are identified and quantified by tandem mass spectrometry combined with database searching. The data are finally analyzed using a suite of software tools ***(11–13)***.

2. Materials

2.1. Isolation of N-Linked Glycopeptides From a Complex Protein Mixture

1. Coupling buffer: 100 m*M* NaOAc, 150 m*M* NaCl (pH 5.5).
2. Sodium periodate: 10.7 mg/mL stock in water (50 m*M* stock solution; light sensitive, make fresh as required).
3. Tube rocker.
4. Affi-gel Hz hydrazide gel (cat. no. 153-6047, Bio-Rad Laboratories).
5. Denaturing buffer: 8 *M* urea, 0.4 *M* NH_4HCO_3; make fresh as required.
6. 80% acetonitrile.
7. 100% methanol.
8. 100 m*M* ammonium bicarbonate (make fresh as required).
9. Milli-Q water or equivalent.
10. PNGase F (P0705S, New England Biolabs, store at 4°C): resuspend 0.3 µL PNGase F in 0.3 mL of 0.1 *M* ammonium bicarbonate buffer (make fresh).
11. Desalting column (732-2010, Bio-Rad Laboratories).
12. Tris(2-carboxyethyl)phosphine hydrochloride (TCEP): make 200 m*M* stock in water (Pierce, Rockford IL).

13. Iodoacetamide: make 46.5 mg/mL stock in water (250 m*M*, make fresh).
14. 1.5 *M* sodium chloride.
15. 0.4% acetic acid.
16. Sequencing-grade trypsin (Promega).
17. SpeedVac.
18. 1-mL glass vial with polyethylene snap cap (WAT025054, Waters).

2.2. N-Terminal Labeling of Glycopeptides

1. Succinic anhydride (Sigma).
2. Succinic-d4-anhydride (C/D/N Isotopes, Pointe-Claire, Quebec, CA).
3. 15% NH_4OH (NH_4OH/H_2O = 15/85 v/v, pH > 11.0).
4. 1 *M* methylisourea in 15% NH_4OH.
5. Dimethylformamide (DMF)/pyridine/H_2O = 50/10/40 (v/v/v).

2.3. Sodium Dodecyl Sulfate-Polyacrylamide Gel Electrophoresis and Detection of Proteins by Silver Staining

1. Control glycoprotein sample: α1-antitrypsin, α2-HS-glycoprotein, and α1-antichymotrypsin (Calbiochem, San Diego, CA).
2. Sodium dodecyl sulfate (SDS)-polyacrylamide gel electrophoresis (PAGE) gel (Bio-Rad).
3. Running buffer: 25 m*M* Tris, 200 m*M* glycine, 0.1% (w/v) SDS.
4. 3X sample loading buffer: 187.5 m*M* Tris-HCl (pH 6.8 at 25°C), 6% w/v SDS, 30% glycerol, 150 m*M* dithiothreitol (DTT), 0.03% w/v bromophenol blue.
5. Prestained molecular-weight markers.
6. Fixation solution: 50% methanol, 10% acetic acid.
7. Silver staining sensitization solution: 0.02% sodium thiosulfate, 0.2 g/L of $Na_2S_2O_3 \cdot H_2O$, freshly made.
8. $AgNO_3$, 0.2 g/100 mL
9. Developing solution (100 mL): 3 g Na_2CO_3, 50 μL of 37% HCOH, 2 mL of 0.02% of sodium thiosulfate.
10. Stopping solution: 14 g/L ethylenediamine tetraacetic acid (EDTA).

2.4. Identification of N-Linked Glycopeptides by Tandem Mass Spectrometry

1. LCQ ion-trap mass spectrometer (Thermo Finnigan, San Jose, CA).
2. Peptide cartridge packed with Magic C18 resin (Michrome Bioresources, Auburn, CA).
3. FAMOS autosampler (DIONEX, Sunnyvale, CA).
4. Microcapillary high-performance liquid chromatography (HPLC) (10 cm × 75 μm i.d.) column packed with Magic C18 resin (Michrome Bioresources, Auburn, CA).
5. HP1100 solvent-delivery system (Hewlett-Packard, Palo Alto, CA).
6. Solvent B: acetonitrile, 0.4% acetic acid, and 0.005% heptafluorobutyric acid (HFBA).
7. Solvent A: 0.4% acetic acid and 0.005% HFBA.

3. Methods

3.1. Isolation of N*-Linked Glycopeptides From a Complex Protein Mixture*

1. Suspend 1 mg of protein in 80 μL of coupling buffer containing 100 m*M* NaOAc, 150 m*M* NaCl (pH 5.5) (*see* **Note 1**).
2. Add 20 μL of sodium periodate stock solution. Mix well and gently shake the suspension using a tube rocker for 60 min at room temperature in the dark.
3. Prepare 500 μL of 50% hydrazide slurry per sample. Wash hydrazide resin three times by adding three column volumes of water and spinning the resin at 3000*g* for 5 min after each wash. Wash/equilibrate three more times with three column volumes of coupling buffer and spin the resin down again. After equilibration, resuspend the hydrazide resin with 250 μL of coupling buffer to make a 50% slurry (*see* **Note 2**).
4. Equilibrate a desalting column twice with 20 mL of coupling buffer for each sample. Add the oxidized protein sample to the prepared desalting columns, and collect the protein using a new collection tube. This will remove excess sodium periodate from the sample (*see* **Note 3**).
5. Save a small amount of the protein before adding the hydrazide resin (*see* **Note 4**).
6. Add hydrazide resin equilibrated in coupling buffer to the sample. Mix well and gently shake the suspension using a tube rocker for 10–24 h at room temperature.
7. Spin the resin down at 1000*g* for 5 min, and save the supernatant after coupling. Proteins remaining in the supernatant can be analyzed by SDS-PAGE and compared with the protein profile before coupling from **step 5** (*see* **Subheading 3.3.** for details).
8. Wash the resin three times with 1 mL of denaturing buffer, followed by spins at 3000*g* for 5 min after each wash. After the last wash and removal of the denaturing buffer, the hydrazide resin is resuspended with 300 μL of denaturing buffer.
9. Add 12 μL of TCEP solution and incubate at room temperature for 30 min.
10. Add 16 μL of prepared iodoacetamide solution to the hydrazide resin and incubate at room temperature for another 30 min.
11. Wash the resin three times with 1 mL of denaturing buffer.
12. After the last wash and removal of the denaturing buffer, the hydrazide resin is resuspended in 50 μL of denaturing buffer. Trypsin is added at 10 μg in 900 μL of water for each sample, and incubated at 37°C overnight.
13. Eliminate nonglycosylated peptides by spinning down at 1000*g* for 5 min and removing the supernatant. This fraction contains the nonglycosylated peptides (*see* **Note 5**).
14. The trypsin-released peptides are removed by washing the resin three times each with 700 μL of the following solutions: 1.5 *M* NaCl, 80% acetonitrile, 100% methanol, water, and ammonium bicarbonate buffer. (For isotopic labeling, proceed to **Subheading 3.2.** after the third methanol wash step.) Remove supernatant by spinning at 3000*g* for 5 min after each wash step.
15. Prepare PNGase F by diluting 0.3 μL of PNGase F with 0.3 mL of ammonium bicarbonate buffer. *N*-linked glycopeptides are released from the resin by the addi-

tion of 0.3 mL of diluted PNGase F solution, followed by incubation at 37°C with mixing overnight.

16. Spin down the resin at 1000*g* for 5 min, transfer the supernatant to a glass vial, and wash the hydrazide resin twice with 200 µL of 80% acetonitrile. Combine the washes with the supernatant in the glass vial.
17. The released glycopeptides are dried in a SpeedVac and resuspended in 10 µL of 0.4% acetic acid for mass-spectrometry analysis (*see* **Subheading 3.3.** for detail; 5 µL can be used in each analysis).

3.2. Isotope Labeling of Glycopeptides

1. For isotopic labeling of glycopeptides with succinic anhydride following the methanol wash at **Subheading 3.1., step 14**, wash the glycopeptide-attached resin twice with 1 mL of 15% NH_4OH (pH > 11.0).
2. Add 200 µL of 1 *M* methylisourea in 15% NH_4OH (NH_4OH/H_2O = 15/85 v/v) and incubate at 55°C for 10 min.
3. Wash the beads twice with 1 mL of water, twice with 1 mL of DMF/pyridine/H_2O = 50/10/40 (v/v/v), and resuspend in 200 µL of DMF/pyridine/H_2O = 50/10/40 (v/v/v).
4. Add the d0 or d4 succinic anhydride solution to a final concentration of 2 mg/mL. Incubate the mixture at room temperature for 1 h.
5. Wash the mixture three times with 1 mL of DMF, three times with 1 mL of water, and six times with 1 mL of 100 m*M* ammonium bicarbonate.
6. Release peptides from the beads using PNGase F as described in **Subheading 3.1., steps 15–17**.

3.3. Determination of Glycopeptide Capture Efficiency From a Three-Control Protein Mix Using SDS-PAGE and Silver Staining

1. Three glycoproteins (50 µg each) are suspended together in 80 µL of water.
2. This control protein sample is then coupled to hydrazide beads as described in **Subheading 3.1., steps 1–7**.
3. Aliquots are removed from the sample both before and after overnight hydrazide resin coupling.
4. Identical aliquots of all samples are mixed with enough 3X sample-loading buffer to yield a 1X concentration (each control protein is also included separately for analysis on SDS-PAGE).
5. Proteins are denatured by heating the samples at 95°C for 5 min, then cooled on ice.
6. Samples are microcentrifuged for 5 min.
7. Proteins are loaded onto an SDS-PAGE gel and separated at 200 V for 1 h, following the manufacturer's instructions for the specific electrophoresis system. SDS-denatured proteins are negatively charged and will migrate to the anode (+).
8. After electrophoresis, the gel is fixed by agitation in a 50% methanol, 10% acetic acid solution for half an hour.
9. The gel is washed five times with deionized water for 20 min each wash.

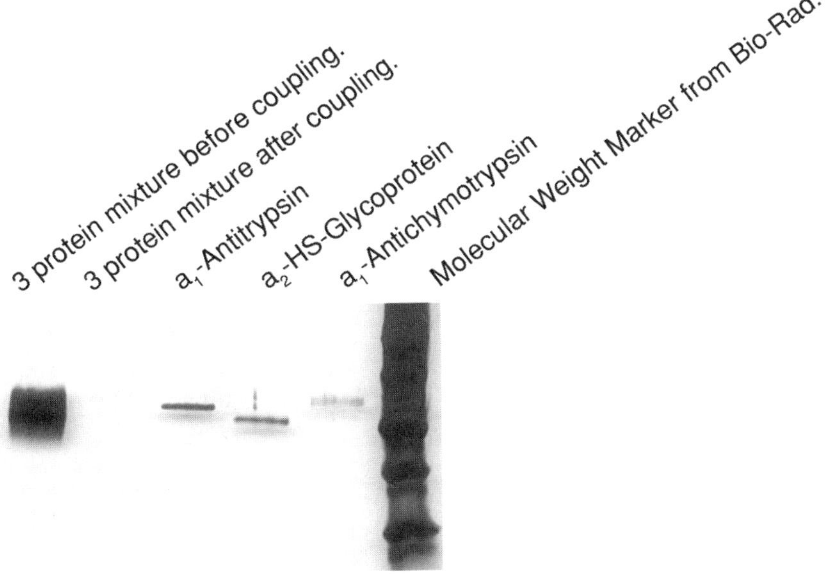

Fig. 2. Capture of glycoproteins by hydrazide resin. Total protein staining of three standard glycoproteins before and after capture of glycoproteins by hydrazide resin. Proteins were separated by sodium dodecyl sulfate-polyacrylamide gel electrophoresis and stained with silver staining reagents.

10. Then the gel is incubated in silver staining sensitization solution ($Na_2S_2O_3$, 0.2 g/L) for 2 min.
11. Following incubation, the gel is washed three times with deionized water for 30 s each wash.
12. The gel is then incubated in an $AgNO_3$ solution (0.2 g/100 mL) for 25 min at 4°C.
13. After incubation, the gel is washed three times with deionized water for 60 s each wash.
14. The gel is incubated with developing solution for up to 10 min. Proteins are visualized as dark brown bands, and longer incubation times will increase the background.
15. The development is stopped by incubating with stopping solution for at least 10 min.
16. Afterwards, the gel is washed with deionized water for a length of time, then stored in water.
17. The gel is dried and the gel image is scanned.

The results are shown in **Fig. 2**. As expected, all three protein bands were essentially depleted by the coupling reaction (silver-stained bands present before and absent after the coupling). This result shows that the hydrazide beads bind the oxidized glycoproteins efficiently.

3.4. Identification of N-Linked Glycopeptides by Tandem Mass Spectrometry

1. The peptides and proteins are identified by tandem mass spectrometry using an LCQ ion-trap mass spectrometer.
2. Formerly *N*-linked glycopeptides are injected into a homemade peptide cartridge packed with Magic C18 resin using a FAMOS autosampler.
3. The peptides are then passed through a 10 cm × 75 μm i.d. microcapillary HPLC column. The C18 peptide trap cartridge, μ-ESI-emitter/μ-LC pulled-tip column combination, high-voltage ESI line, and waste line are each connected to separate ports of a four-port union (Upchurch Scientific, Oak Harbor, WA) constructed entirely of polyetheretherketone ***(14)***.
4. A linear gradient of acetonitrile with 5–32% solvent B over 100 min at a flow rate of approx 300 nL/min is applied for reverse-phase liquid chromatography using an HP1100 solvent-delivery system.
5. The eluting peptides are directly ionized by electrospray ionization and detected, and the specific peptide ions are automatically selected and fragmented by collision-induced dissociation (CID) to generate a tandem mass spectrum (MS/MS spectrum or CID mass spectrum). The mass spectrometer is operated in a data-dependent MS/MS mode (a full-scan mass spectrum is followed by a tandem mass spectrum), where the precursor ion is selected "on the fly" from the previous scan. An *m/z* ratio for an ion that has been selected for fragmentation is placed in a list and dynamically excluded for 3 min from further fragmentation.
6. The acquired MS/MS spectra are searched against the human protein database using the computer program SEQUEST ***(11)***. The mass window for each peptide being searched is given a tolerance of 3 Da between the measured average mass and the calculated average mass, with the b and y ion series included in the SEQUEST analysis. The sequence database search is set to expect the following modifications: carboxymethylated cysteines, oxidized methionines, and an enzyme-catalyzed conversion of Asn to Asp at the site of carbohydrate attachment. There are no other constraints included in the SEQUEST search.

4. Notes

1. Proteins in other solutions can be transferred to coupling buffer by buffer exchange.
2. Other hydrazide resins can be used to couple the oxidized glycoproteins to a solid support. The capacity and the amount of hydrazide resin needed for efficient coupling of proteins should be determined using the protein standards described under **Subheading 3.3.**
3. Buffer exchange or removal of excess periodate can also be achieved by protein precipitation, protein dialysis against the coupling buffer, or by sample filtration using filters with a low-molecular-weight cutoff.
4. Sample aliquots removed for SDS-PAGE analysis both before and after hydrazide coupling should be adjusted for any volume changes due to the addition of hydrazide resin.
5. The trypsin-released peptides are removed and collected. The peptides can be analyzed to identify peptides from proteins containing a single *N*-linked glycosylated peptide or no *N*-linked glycosylated peptide.

References

1. Roth, J. (2002) Protein *N*-glycosylation along the secretory pathway: relationship to organelle topography and function, protein quality control, and cell interactions. *Chem. Rev.* **102,** 285–303.
2. Varki, A. E. A. (1999) *Essentials of Glycobiology*. Cold Spring Harbor Laboratory Press, Cold Spring Harbor, NY
3. Bause, E. (1983) Structural requirements of *N*-glycosylation of proteins. Studies with proline peptides as conformational probes. *Biochem. J.* **209,** 331–336.
4. Diamandis, E. P. (2004) Mass spectrometry as a diagnostic and a cancer biomarker discovery tool: opportunities and potential limitations. *Mol. Cell. Proteomics* **3,** 367–378.
5. Shepard, H. M., Lewis, G. D., Sarup, J. C., et al. (1991) Monoclonal antibody therapy of human cancer: taking the HER2 protooncogene to the clinic. *J. Clin. Immunol.* **11,** 117–127.
6. Durand, G. and Seta, N. (2000) Protein glycosylation and diseases: blood and urinary oligosaccharides as markers for diagnosis and therapeutic monitoring. *Clin. Chem.* **46,** 795–805.
7. Zhang, H., Yan, W., and Aebersold, R. (2004) Chemical probes and tandem mass spectrometry: a strategy for the quantitative analysis of proteomes and subproteomes. *Curr. Opin. Chem. Biol.* **8,** 66–75.
8. Zhang, H., Li, X. J., Martin, D. B., and Aebersold, R. (2003) Identification and quantification of *N*-linked glycoproteins using hydrazide chemistry, stable isotope labeling and mass spectrometry. *Nat. Biotechnol.* **21,** 660–666.
9. Zhang, H., Yi, E. C., Li, X. J., et al. (2005) High throughput quantitative analysis of serum proteins using glycopeptide capture and liquid chromatography mass spectrometry. *Mol. Cell. Proteomics* **4,** 144–155.
10. Pan, S., Zhang, H., Rush, J., et al. (2005) High throughput proteome screening for biomarker detection. *Mol. Cell. Proteomics* **4,** 182–190.
11. Eng, J., McCormack, A. L., and Yates, J. R., 3rd. (1994) An approach to correlate tandem mass spectral data of peptides with amino acid sequences in a protein database. *J. Am. Soc. Mass Spectrom.* **5,** 976–989.
12. Han, D. K., Eng, J., Zhou, H., and Aebersold, R. (2001) Quantitative profiling of differentiation-induced microsomal proteins using isotope-coded affinity tags and mass spectrometry. *Nat. Biotechnol.* **19,** 946–951.
13. Keller, A., Nesvizhskii, A. I., Kolker, E., and Aebersold, R. (2002) Empirical statistical model to estimate the accuracy of peptide identifications made by MS/MS and database search. *Anal. Chem.* **74,** 5383–5392.
14. Yi, E. C., Lee, H., Aebersold, R., and Goodlett, D. R. (2003) A microcapillary trap cartridge-microcapillary high-performance liquid chromatography electrospray ionization emitter device capable of peptide tandem mass spectrometry at the attomole level on an ion trap mass spectrometer with automated routine operation. *Rapid Commun. Mass Spectrom.* **17,** 2093–2098.

13

N-Glycosylation Analysis Using the StrOligo Algorithm

Martin Ethier, Daniel Figeys, and Hélène Perreault

Summary

N-glycosylation of proteins is the predominant glycosylation in mammals and confers specific conformations, localization, and functions to proteins. High-throughput proteomics techniques have focused on the identification of proteins through amino acid sequence determination, with little attention paid to their post-translational modification, in particular, glycosylation. High-throughput mass spectrometric data often contain information about glycosylation, but this is systematically discarded by proteomic search engines. We have developed an algorithm, StrOligo (for STRucture of OLIGOsaccharides), capable of automated analysis of oligosaccharide composition and possible structures by mass spectrometry.

The algorithm analyzes tandem mass spectrometry (MS/MS) data in an automated three-step process and provides possible structures and a discrimination score. In the first step, the algorithm constructs a relationship tree of the monosaccharide moiety losses observed in the MS/MS spectrum. In the second step, the algorithm uses the tree to propose possible compositions and structures from combinations of adduct and fragment ions as well as a discrimination score, which reflects the fit with the experimental results. Finally, an interface is available to visualize the proposed structures and their scores. As well, the MS/MS spectrum is displayed with relevant peaks labeled for the proposed structure with the highest discrimination score, using a modified nomenclature.

Key Words: *N*-linked oligosaccharides; mammalian cells; structure determination; tandem mass spectrometry; automated interpretation; algorithm; sugar composition.

1. Introduction

The advent of high-throughput technologies in genomics and proteomics has led the way to development of bioanalytical methods involving automated instrumentation and interpretation of data. Matrix-assisted laser desorption/ionization (MALDI)-mass spectrometry (MS) is now recognized as a technique of primary importance as part of automated bioanalytical systems, because of its simplicity and rapidity of data acquisition. The production of predominant singly charged $[M+H]^+$ and $[M+Na]^+$ ions of biomolecules and the ability to

From: *Methods in Molecular Biology, Vol. 328: New and Emerging Proteomic Techniques*
Edited by: D. Nedelkov and R. W. Nelson © Humana Press Inc., Totowa, NJ

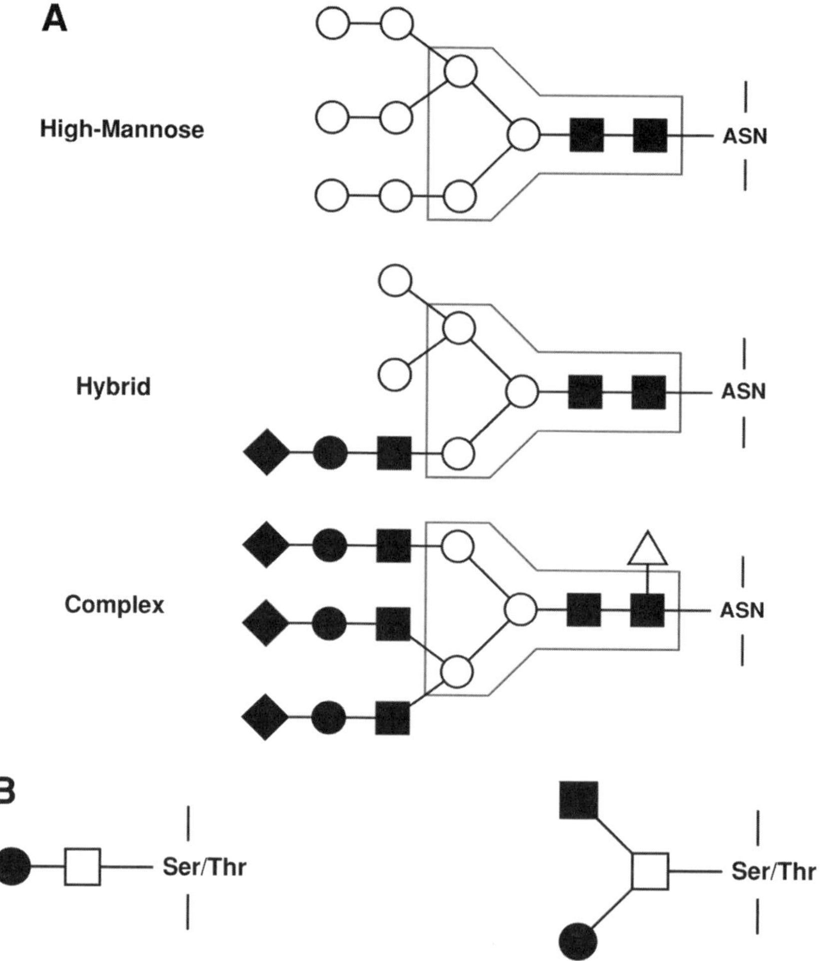

Fig. 1. Type of oligosaccharides attached to proteins **(A)** *N*-linked glycans and **(B)** *O*-linked glycans. Solid circle, galactose; open circle, mannose; solid square, *N*-acetyl-glucosamine; open triangle, fucose; solid diamond, sialic acid.

elucidate the structures of such MALDI ions using either post-source decay (PSD) *(1)* or tandem MS (MS/MS) *(2,3)* have made this technique highly useful in bioanalysis.

The elucidation of glycan structure and function in glycoproteins remains one of the most challenging tasks given to biochemists and bioanalysts, and constitutes a very important field of posttranslational modification studies. Although some glycosylation rules are already established according to biosynthetic pathways, several types of carbohydrates (**Fig. 1**) have been characterized in glycoproteins *(4)*, and thus rules are very particular to each type considered. *N*-linked glycans in mammals are beta-linked to the amide nitrogen of asparagine (Asn) in the consensus sequence Asn-X-Ser/Cys/Thr, where X can be any amino acid except proline. These glycans have in common a trimannosyl core, which also comprises two *N*-acetylglucosamine residues. Depending on the type of branching originating from the core, they are classified as high-mannose, hybrid, or complex structures. These *N*-linked structures are thus predictable, whereas in the case of *O*-glycosylation, no simple motif can predict sugar structures. *O*-glycosylation occurs on the hydroxyl group of, for example, threonine, tyrosine, and serine, and gives rise to diverse glycan types.

Recent literature shows that efforts have been made to automate the interpretation of tandem mass spectra of oligosaccharides *(5–7)*; however, each technique is limited to one part of the analysis.

The StrOligo algorithm *(8–10)* described in this study uses some routines, although modified, from three existing studies *(5–7)* and joins them in a new innovative fashion to perform the analysis of *N*-linked glycans. User intervention is kept to a minimum. It must be noted that although this algorithm considerably reduces the time needed for interpretation, it is not intended to be a complete replacement for human intervention. It is intended to quickly suggest possible structures to facilitate the analysis and direct the user towards further adequate experiments.

2. Materials

1. Sample: Recombinant human β-interferon (β-IFN) extracted from a Chinese hamster ovary (CHO) cell line stably transfected with human β-IFN gene.
2. Sample preparation:
   a. Deglycosylation was performed using PNGase F enzyme (Roche, Indianapolis, IN).
   b. Labeling of the glycans was performed using 2-aminobenzamide (2-AB) (Sigma-Aldritch, St. Louis, MO).
   c. Sample was deposited using the dried-drop method on the MALDI target using 2,5-dihydroxybenzoic acid (DHB) (Sigma-Aldrich, St. Louis, MO) as the matrix.
3. Tandem mass spectrometry (MS/MS): MALDI-MS/MS spectra were acquired on a prototype hybrid quadrupole-quadrupole-time-of-flight (QqTOF) mass spectrometer, built in the Department of Physics and Astronomy, University of Manitoba *(11)*.

3. Methods

The StrOligo algorithm is composed of four main steps: isotopic simplification, relationship tree building, composition analysis, and proposition of structures.

3.1. Isotopic Simplification

Biomolecules contain large numbers of atoms. Because most atoms possess different isotopes, an accumulation of atoms will result in a complicated isotopic distribution for the molecule. Using the QqTOF instrument, it is possible to differentiate some of the isotopes owing to sufficient resolving power (approx 10,000). The mass spectrum features a peak representing the monoisotopic state (all atoms are in their lightest isotope state) and other peaks corresponding to the addition of one or more neutrons to any atom. The height of a peak in an isotopic cluster in mass spectrometry is dictated by the number of atoms of a given type and the probability of appearance of all isotopes of that atom.

Combinatorial mathematics can then be used to predict the appearance of the isotopic distribution in the mass spectrum. With increasing molecular size, the isotopic distribution becomes more and more complicated. The atoms contained in nonsulfated glycans are usually limited to C, H, N, and O, and the probability of occurrence of their isotopes is well characterized.

In an automated analysis, every peak is analyzed. One must take isotopic distributions into account or the analysis will be complicated by superfluous information from these distributions. Addition of every peak from isotopic distributions at the monoisotopic peak positions would solve this problem. Because the appearance of isotopic distribution changes with increasing mass, no simple pattern can fit all distributions throughout the mass spectrum in order to simplify all isotope clusters to a single monoisotopic peak.

Calculating the predicted isotopic distribution for each peak would solve the problem of relative intensities varying with increasing *m/z*. To do so, an empirical formula would be needed for each peak. Such information is not readily available, but knowledge about the type of molecules studied can help. Oligosaccharides are composed of monosaccharide units that all have atomic compositions approaching a $C_nH_{2n+1}O_n$ ratio. By calculating an average mass for a monosaccharide unit based on experimental evidence, it is possible to find an approximate empirical formula for any mass value.

Monoisotopic peaks are thus identified by calculating a predicted isotopic distribution for any *m/z* value for an oligosaccharide.

1. The file containing the MS/MS spectrum to analyze is opened using the StrOligo algorithm. The data from the mass spectrometer is fed to the algorithm as an ASCII format text file (flat file) (*see* **Note 1**).
2. The parent ion *m/z* value used to acquire the MS/MS spectrum is written in the parent ion monoisotopic mass field. **Figure 2** shows the information that the algorithm needs.
3. The ion mode in which the acquisition was made is also provided to the algorithm.
4. The labeling agent used to derivatize the glycans is selected from the list of available labels (*see* **Note 2**).

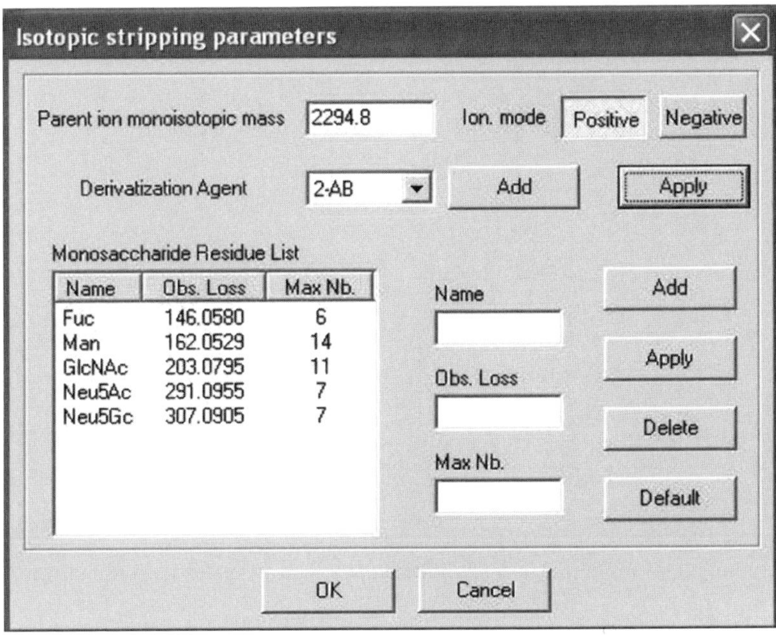

Fig. 2. Windows used to gather information about the experiment.

5. By clicking on the Apply button, the algorithm calculates the maximum number of monosaccharides of each type allowed by the parent ion mass entered (*see* **Note 3**).
6. By clicking on the OK button, the software opens the MS/MS spectrum and shows it to the user (*see* **Note 4**).
7. The user can perform the isotopic simplification by clicking on the S icon. **Fig. 3** shows an example of simplified MS/MS spectrum.

3.2. Relation Tree

Tandem mass spectra contain many types of peaks. Some are due to fragments of the parent ion and others are due to fragments of contaminating species or background signal. Only the fragment peaks deriving from the parent ion are informative. The algorithm also analyzes superfluous peaks as potential fragments, thus increasing analysis time and the chance of random fitting. It is thus necessary to discard these uninformative peaks.

Some fragments observed, called inter-ring fragments, arise from the cleavage of the molecule between two monosaccharides. Other fragments, called cross-ring fragments, originate from the cleavage in the middle of a monosaccharide. Depending where the cleavage occurs in the monosaccharide, cross-ring cleavages can give a variety of fragments. Inter-ring fragments have fewer possibilities of cleavage positions and are thus easier to explain. The goal of this

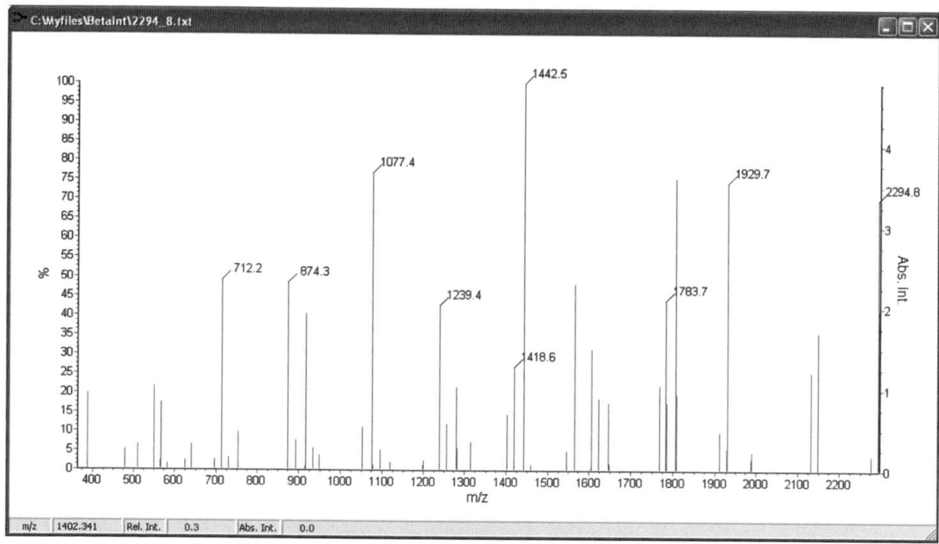

Fig. 3. Example of tandem mass spectrometry spectrum after isotopic simplification.

algorithm was to study the monosaccharide's sequence; thus, only inter-ring fragments are necessary. Cross-ring fragments would be needed only for specific connectivity analysis.

Similarly to the method used in protein sequence analysis by mass spectrometry ***(12)***, the identity of a residue (either an amino acid or a monosaccharide) can be determined by finding the *m/z* value between two fragments differing by only one residue. Because only inter-ring fragments are monitored in this analysis, only the loss of whole monosaccharides will be considered. Thus, fragments differing by only one residue should be easy to find, and identification of monosaccharide residues present should be possible.

The information gathered from this analysis is presented to the user in a graphical form called a relationship tree. The tree consists of related nodes representing peaks. Peaks are related if the difference in *m/z* corresponds to the loss of a known residue, thus allowing the identification of one monosaccharide present in the glycan.

To build the relationship tree, the algorithm looks for losses of intact monosaccharides in the mass spectrum. The difference of *m/z* in each pair of peaks is compared to a list of losses of known monosaccharide (*see* **Note 5**).

1. While the simplified MS/MS spectrum is the active window, the user can make the algorithm build the relationship tree by clicking on the RT icon.
2. The user then needs to provide the algorithm with a threshold value. Peaks with heights under the threshold will be neglected (*see* **Note 6**).

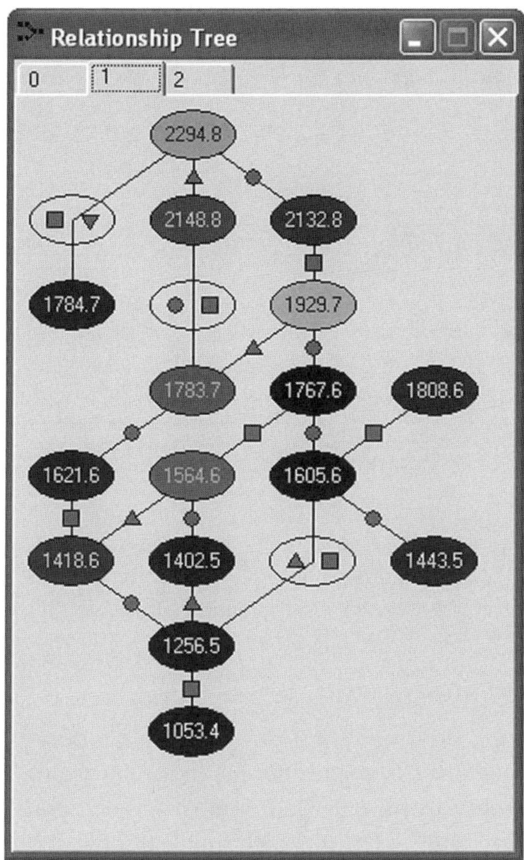

Fig. 4. Example of relationship tree built.

3. The user can also provide the tree with a degree of optimization (*see* **Note 7**).
4. The tree is composed of nodes, which represents peaks. The links between nodes correspond to known losses. The different tabs represent series of peaks that could not be related by the algorithm. **Figure 4** shows a typical relationship tree obtained by this method.

3.3. Composition Analysis

The next step is to find the monosaccharide composition of the glycan. The algorithm starts by finding probable compositions. To achieve this, the algorithm finds every possible combination of monosaccharides totaling to the *m/z* value of the glycan studied. A score is assigned to each composition based on its agreement with the relationship tree.

1. While the relationship tree is the active window, the user can start the composition analysis by clicking on the FC icon. **Figure 5** shows a typical composition analysis.

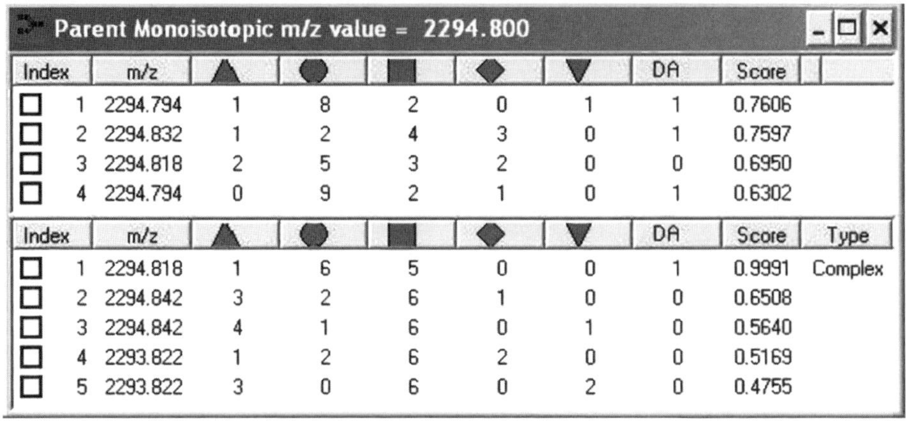

Parent Monoisotopic m/z value = 2294.800

Index	m/z	▲	●	■	◆	▼	DA	Score
☐ 1	2294.794	1	8	2	0	1	1	0.7606
☐ 2	2294.832	1	2	4	3	0	1	0.7597
☐ 3	2294.818	2	5	3	2	0	0	0.6950
☐ 4	2294.794	0	9	2	1	0	1	0.6302

Index	m/z	▲	●	■	◆	▼	DA	Score	Type
☐ 1	2294.818	1	6	5	0	0	1	0.9991	Complex
☐ 2	2294.842	3	2	6	1	0	0	0.6508	
☐ 3	2294.842	4	1	6	0	1	0	0.5640	
☐ 4	2293.822	1	2	6	2	0	0	0.5169	
☐ 5	2293.822	3	0	6	0	2	0	0.4755	

Fig. 5. Example of composition analysis.

2. The user is presented with two sets of compositions. The top set is for compositions with an H^+ adduct and the bottom set is for the Na^+ adduct (*see* **Note 8**).
3. The user can verify the extent of agreement by clicking on the index of a composition (*see* **Note 9**).

3.4. Proposition of Structures

The last step of the StrOligo algorithm is to assign possible structures to the molecule represented by the experimental mass spectrum. StrOligo starts by finding possible structures for a given composition using accepted biosynthetic rules for *N*-linked glycans. A template summarizing these rules was created and included in the algorithm.

To suggest a possible structure, StrOligo builds possible structures according to the template using one monosaccharide at a time, taken from the composition of interest, until every monosaccharide is accounted for. A score is then assigned to each possibility based on the comparison of a theoretical MS/MS spectrum to the experimental data (*see* **Note 10**).

1. While composition analysis is the active window, the user can start the structural analysis by clicking on the FAS icon. Each composition is then tested for possible structures. **Figure 6** shows a typical structural analysis.
2. The different tabs represent possible structures. The number on the tab is the score. A higher score corresponds to a better match.
3. When a tab is clicked on, peaks explained in the simplified MS/MS spectrum are automatically labeled with the corresponding fragmentation nomenclature (*see* **Note 11**).

4. Notes

1. The ASCII files consists of a list of *m/z* values and intensities. Each line contains only one *m/z* value and its corresponding intensity separated by a tab. The file is

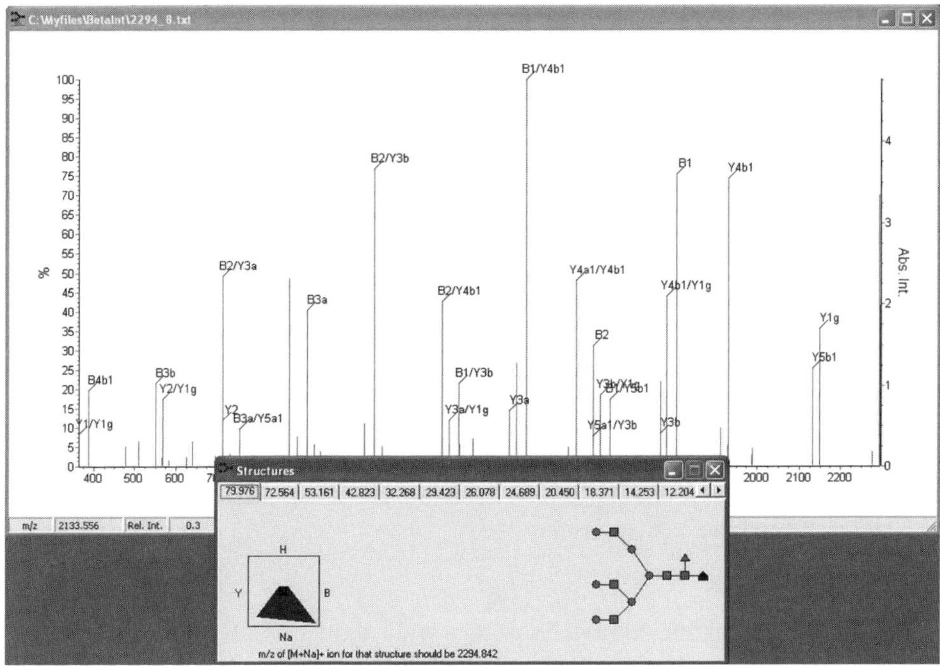

Fig. 6. Example of structural analysis with automated labeling of fragments.

sorted in increasing values of *m/z*. The transfer of the data to this format is highly dependent on the acquisition software. No attempt to cover these details will be made here.

2. If the label used is not in the algorithm, it is possible to add it to the list by providing the algorithm with the number of each atom type added to the glycan and the maximum number of label(s) added to the glycan.
3. Detailed structural analysis is limited to the five most commonly encountered monosaccharides. Other monosaccharides can be added to the list by providing the algorithm with a name, observed loss in the mass spectrometer, and the maximum number of units that can be added to the glycan. However, manually added monosaccharides allow only partial analysis, because the biosynthetic pathways for these monosaccharides are not included in the original algorithm. In these instances, the analysis will stop at the composition step.
4. The user can zoom in on a desired region by clicking and holding on the left mouse button on the upper-left corner of the region and dragging the mouse to the lower-right corner of the region. To come back to the full spectrum, the user must click and hold the left mouse button and drag the mouse toward the upper-left corner.
5. Because experimental variations in measured *m/z* values are expected, peaks falling within a window of 50 mDa around the value of the known loss are accepted.

6. The user will choose a value that rejects most of the noise but will keep most of the valid peaks. A typical value of five is suggested, but can be adjusted depending on the quality of the spectrum.
7. The degree of optimization has nothing to do with the analysis. It is used to improve the visual appearance of the tree. If the user is interested in a quick analysis, it can set the degree to "none."
8. The type of *N*-linked glycan present is indicated beside the composition. Only complex *N*-linked glycans can figure as proposed structures.
9. The color of the nodes in the relationship tree will be change to reflect this. Red nodes are in agreement, whereas white ones are not.
10. The algorithm calculates a score for each possibility of adduct and each type of fragment. The global figure of merit is represented as a geometrical shape. The larger its area, the better the agreement.
11. The nomenclature used is described in **ref. *9***. Y fragments correspond to ions with the charge on the reducing end. B fragments are for ions with the charge on the other end. Indices represent the number of monosaccharides between the fragmentation site and the reducing end.

Acknowledgments

The authors thank the National Sciences and Engineering Research Council of Canada (NSERC) for funds. Acknowledgments are also directed to the Canada Research Chair Program (H. P. and D. F.). The Canadian Foundation for Innovation (CFI) is also thanked. Martin Ethier is the recipient of an NSERC postgraduate fellowship. The integrin sample was provided by Dr. M. Butler, Microbiology, University of Manitoba. The spectra were acquired by Dr. Oleg Krokhin on the MALDI-QqTOF mass spectrometer built in the laboratory of Drs. W. Ens and K. G. Standing, Physics and Astronomy, University of Manitoba.

References

1. Viseux, N., Costello, C. E., and Domon, B. J. (1999) Post-source decay mass spectrometry: optimized calibration procedure and structural characterization of permethylated oligosaccharides. *J. Mass Spectrom.* **34,** 364–376.
2. Harvey, D. J., Bateman, R. H., Bordoli, R. S., and Tyldesley, R. (2000) Ionisation and fragmentation of complex glycans with a quadrupole time-of-flight mass spectrometer fitted with a matrix-assisted laser desorption/ionisation ion source. *Rapid Commun. Mass Spectrom.* **14,** 2135–2142.
3. Saba, J. A., Kunkel, J. P., Jan, D. C. H., et al. (2002) A study of immunoglobulin G glycosylation in monoclonal and polyclonal species by electrospray and matrix-assisted laser desorption/ionization mass spectrometry. *Anal. Biochem.* **305,** 16–31.
4. Varki, A., Cummings, R., Esko, J., Freeze, H., Hart, G., and Marth, J. (eds) (1999) *Essentials of Glycobiology.* Cold Harbor Laboratory, Cold Spring Harbor, NY.

5. Mizuno, Y., Sasagawa, T., Dohmae, N., and Takio, K. (1999) An automated interpretation of MALDI/TOF postsource decay spectra of oligosaccharides. 1. Automated peak assignment. *Anal. Chem.* **71,** 4764–4771.
6. Cooper, C. A., Gasteiger, E., and Packer, N. H. (2001) GlycoMod-a software tool for determining glycosylation compositions from mass spectrometric data. *Proteomics* **1,** 340–349.
7. Gaucher, S. P., Morrow, J., and Leary, J. A. (2000) STAT: a saccharide topology analysis tool used in combination with tandem mass spectrometry. *Anal. Chem.* **72,** 2331–2336.
8. Ethier, M., Saba, J. A., Ens, W., Standing, K. G., and Perreault, H. (2002) Automated structural assignment of derivatized complex *N*-linked oligosaccharides from tandem mass spectra. *Rapid Commun. Mass Spectrom.* **16,** 1743–1754.
9. Ethier, M., Saba, J. A., Spearman, M., et al. (2003) Application of the StrOligo algorithm for the automated structure assignment of complex *N*-linked glycans from glycoproteins using tandem mass spectrometry. *Rapid Commun. Mass Spectrom.* **17,** 2713–2720.
10. Ethier, M., Krokhin, O., Ens, W., Standing, K. G., Wilkins, J., and Perreault, H. (2005) Global and site-specific detection of human integrin alpha 5 beta 1 glycosylation using tandem mass spectrometry and the StrOligo algorithm. *Rapid Commun. Mass Spectrom.* **19,** 721–727.
11. Loboda, A. V., Krutchinsky, A. N., Bromirski, M. P., Ens, W., and Standing, K. G. (2000) A tandem quadrupole/time-of-flight mass spectrometer with a matrix-assisted laser desorption/ionization source: design and performance. *Rapid Commun. Mass Spectrom.* **14,** 1047–1057.
12. Biemann, K. and Martin, S. A. (1987) Mass spectrometric determination of the amino acid sequence of peptides and proteins. *Mass Spectrom. Rev.* **6,** 1–75.

14

MALDI-MS Data Analysis for Disease Biomarker Discovery

Weichuan Yu, Baolin Wu, Junfeng Liu, Xiaoye Li, Kathy Stone, Kenneth R. Williams, and Hongyu Zhao

Summary

In this chapter, we address the issue of matrix-assisted laser desorption/ionization mass spectrometry (MS) data analysis for disease biomarker discovery. We first give a general framework of MS data analysis, then focus on several key steps. After that, we show some application examples using an ovarian sera cancer dataset. Finally, we discuss the limitations of current approaches and possible future research directions.

Key Words: Proteomics; MALDI-MS data analysis; peak detection; peak alignment; feature selection; sample classification.

1. Introduction

Matrix-assisted laser desorption/ionization (MALDI) mass spectrometry (MS) is widely used in proteomics research as a high-throughput, high-sensitivity technique to identify proteins and their post-translational modifications. In addition, MALDI-MS produces single charges on most peptide/protein ions, thus simplifying the interpretation of the resulting data. In this chapter, we address issues related to the analysis of MALDI-MS data, aiming at developing a robust and efficient biomarker discovery methodology.

Generally, we follow the framework outlined in **Fig. 1** for biomarker analysis of MALDI-MS spectra. First, baselines are corrected to reduce the influence of background noise. Then, the data are smoothed to remove high-frequency noise. After that, biologically significant peaks are identified and aligned together across all sample spectra. These aligned peaks serve as the feature points for subsequent statistical analysis.

Among the previously mentioned steps, baseline correction and smoothing can generally be handled using well-established methods (e.g., during baseline

From: *Methods in Molecular Biology, Vol. 328: New and Emerging Proteomic Techniques*
Edited by: D. Nedelkov and R. W. Nelson © Humana Press Inc., Totowa, NJ

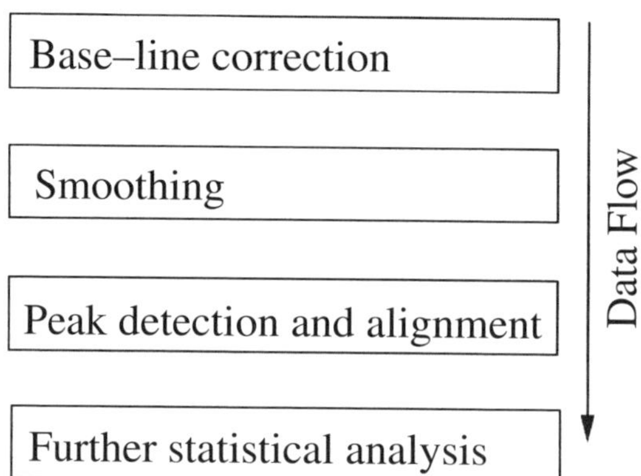

Fig. 1. General framework of biomarker analysis of mass spectrometry data.

correction we usually use curve-fitting techniques to eliminate the influence of background intensity variation before detecting and analyzing peaks *[1]*), whereas approaches for the peak identification and alignment steps that occur during data preprocessing are less well established. Finally, feature selection and classification are the critical last steps that occur during biomarker discovery. In the following, we will address these issues in more detail. Before our general discussion, the critical need for developing statistical as well as visualization tools for biomarker studies is demonstrated in **Fig. 2**. The five *m/z* ratios indicated in this figure by vertical lines were identified by computational analysis in *(2)* as biomarkers of ovarian cancer. However, a closer look at representative mass spectra around these biomarkers strongly suggests they do not arise from the ionization of biologically meaningful peptides that are differentially expressed in sera from normal vs ovarian cancer patients. Rather, the positions of these biomarkers seem to mark regions with differing baseline and/or noise intensities.

2. Materials

The MALDI-MS spectra used in our research were acquired on desalted human sera samples by a Micromass M@LDI-L/R instrument. First, 2 μL of each serum sample is deposited into a 96-well plate (ABgene Thermo-Fast 96,

→

Fig. 2. Five vertical lines denote the location of five biomarkers that were identified in the ovarian cancer example in **ref.** ***17***. Clearly, none of these biomarkers appear to correspond to biologically relevant peaks. The spectra were downloaded from the National Institutes of Health Clinical Proteomics Program Data Bank, http://clinical-proteomics.steem.com/, which is no longer available.

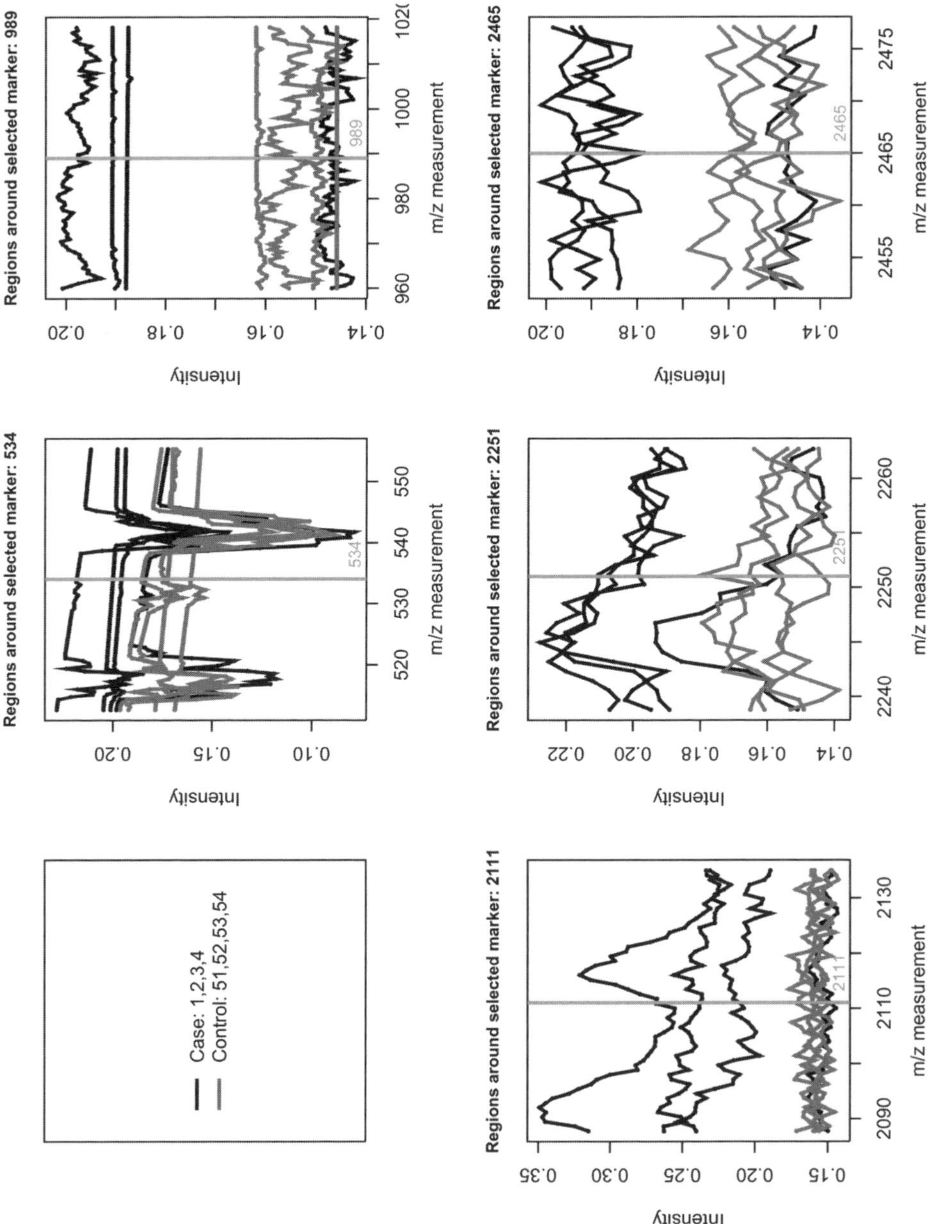
Case: 1,2,3,4
Control: 51,52,53,54
Regions around selected marker: 534
Regions around selected marker: 989
Regions around selected marker: 2111
Regions around selected marker: 2251
Regions around selected marker: 2465
Intensity
m/z measurement

Skirted) and acidified by the addition of 10 μL 0.1% trifluoroacetic acid (TFA; Pierce). A Packard MassPrep robot then desalts the samples in groups of four. The C-18 ZipTips (P10 size, 0.6 μL, Millipore) are washed with 50% acetonitrile (Baker, HPLC grade), 0.05% TFA, and are then equilibrated with 3 × 20 μL 0.1% TFA. Each sample is aspirated up and then expelled back into the original sample well eight times, and the ZipTip is then washed five times with 20 μL 0.1% TFA. Peptides/proteins are eluted from the ZipTip with 10 μL of 50% acetonitrile, 0.1% formic acid, and deposited into a new 96-well plate. A 2-μL aliquot of each sample eluent is manually removed (which was found to be more reliable and efficient than robotic spotting on the Packard robot), mixed with 1.0 μL of a saturated (approx 4.5 mg/mL) solution of alpha-cyano-4-hydroxycinnamic acid matrix (Waters) in 50% acetonitrile, 0.05% TFA, containing an internal standard of 25 fmol/2 μL bradykinin (Sigma) (M+H C12 mono-isotopic mass: 1060.569), as a marker for the reflectron analysis, and 100 fmol/2 μL of insulin chain B, oxidized (Sigma) (M+H average mass: 3496.95) as a linear marker. This sample solution was then spotted manually onto the MALDI-MS target and subjected to automated MALDI-MS on a Micromass M@LDI-L/R mass spectrometer.

The M@LDI-L/R mass spectrometer automatically acquires two sets of data in positive ion detection mode. The mass range acquired is dependent on the mass analyzer being used, with 700–3500 Da for reflectron and 3450–28,000 Da for linear. Although the mass range is adjustable, meaningful data are not acquired below approx 700 Da, because of interference from the matrix. The mass of 28,000 Da is about the upper limit for the α-cyano-4-hydroxy cinnamic acid matrix. For higher masses, the sample would have to be re-spotted using a different matrix. The linear analyzer is used for the high-mass region because with the reflectron analyzer the sensitivity of detection decreases substantially above 3500 Da. Currently, with the laser operating at 5 Hz, the M@LDI-L/R sums 10 individual laser shots to form one spectrum. The laser moves in a random walk around the target well, acquiring data from a maximum of 20 different locations within each 2-mm-diameter well. A spectrum is considered acceptable if it has a signal that is more than 2% above background noise, less than 95% of saturation, and if there is at least one peak detected in the mass range from 1125 Da to 3500 Da for reflectron mode, and one peak between 3450 Da and 28,000 Da for linear mode. The M@LDI-L/R is programmed to retain up to 40 acceptable spectra, but if it sequentially acquires 4 unacceptable spectra, it will move to another location within the same target well. The instrument uses incrementally increasing laser energy to acquire acceptable spectra. This ensures that the lowest possible laser energy is used to obtain acceptable spectra while maintaining the best possible mass resolution. If the M@LDI-L/R acquires 20 acceptable spectra at one position, it will then move to another position in the same sample well, and will acquire another 20 acceptable spectra, unless interrupted by 4 unacceptable spectra. Once the M@LDI-L/R has shot (not acquired) 40 acceptable spectra, it will move to the next sample well. This means there can be a maximum of 40 acceptable spectra acquired for each sample, and that if at no point it acquires acceptable data, it will try up to 10 different locations within the same sample target well before moving on to the next sample. Typically, the resulting spectrum represents the average of 20–40 spectra. The expected mass resolution is 14,000 at M+H 2465, and mass accuracy is better than +70 ppm.

Linear spectra are calibrated using as external calibrants the three peaks derived from equine cytochrome C (M+2H, 6181.05; M+H, 12361.1; and 2M+H, 24721.2). The calibrant is located on target spots adjacent to the sample spots. For data processing, the linear spectra are aligned using the insulin chain B peak/mass.

Each (averaged) reflectron MALDI-MS spectrum is converted to a text-file listing of approx 94,900 *m/z* vs intensity data points spanning the *m/z* range from 700 to 3,500, which is then suitable for further analysis. Data points in the linear MALDI-MS that overlap with the 3450 to 3500 region of the reflectron spectrum are deleted, resulting in a "continuous" spectrum from 700 to 28,000 daltons, with the reflectron/linear breakpoint appearing at 3500 Da. The linear spectrum from 3500 to 28,000 Da contains approx 36,900 data points. The merged reflectron plus linear spectrum thus contains approx 131,800 data points.

All spectra are visually screened so that poor quality spectra can then be reshot manually from the same targets. If the manually acquired spectrum appears to be visually superior to the automatically acquired spectrum, then the file is overwritten with the manually acquired spectrum. Additional information on both automated desalting of serum samples and MALDI-MS data acquisition can be found at http://info.med.yale.edu/wmkeck/prochem/biomarker.htm.

3. Methods

3.1. Data Preprocessing

1. Global alignment: to circumvent the slight variation in *m/z* values for the same relative data points in different spectra, reflectron data points are numbered consecutively by assigning the observed *m/z* value that is closest to the expected MH^+ for the C12 isotope of bradykinin, which is 1060.569, as data point zero. Linear data points are numbered consecutively by assigning the observed *m/z* value that is closest to the expected average MH^+ of oxidized insulin chain B, which is 3496.95, as data point 100,000 (*see* **Notes 1** and **2** for related approaches and our ongoing effort on multiple-peak alignment).
2. Exclusion of data around bradykinin internal standard: To eliminate the bradykinin peak from consideration as a possible marker and to prevent its intensity from

impacting on normalization of reflectron mass spectra (discussed later), *m/z* values that range from 1060.4 to 1063, which include the C12 and the first two of three apparently observable isotope peaks, are excluded from consideration. The third isotope peak potentially could be erroneously determined to be a marker, which is a possibility that needs to be kept in mind. However, by excluding only up to 1063 we leave open the possibility of finding a real marker in the 1063 to 1064 region. Assuming that we exclude the 1060.4 to 1063 region and that data points are numbered as indicated above, then data points numbered from approx -6 to +112 will be excluded from consideration as possible markers.

3. Initially, all isotope peaks from each peptide ion are included in the search for biomarkers. This approach will better allow peptide biomarkers to be identified whose isotopic envelopes overlap with envelopes resulting from nonbiomarker peptides. Hence, if we want the program to search for 10 peptide markers, we need to allow 40 data point markers, with one for each of the four isotopes that might commonly be observed from reflectron data (*see* **Notes 3–7** for our current approaches on envelope extraction).
4. Baseline correction is carried out separately for linear and reflectron data as follows:
   a. Take the natural log of all intensities.
   b. Use a sliding window of 1000 data points; determine the least-squares, robust local polynomial fit curve that best represents the baseline, and for each data point subtract the corresponding baseline intensity.
5. Link together each reflectron spectrum with its corresponding linear spectrum at 3500 Da to produce a spectrum from 700 to 28,000 Da, and exclude the linear internal standard of oxidized B-chain of insulin.
6. Linearly normalize each spectrum and delete unacceptable spectra.
   a. Determine an overall, baseline-corrected median spectrum.
   b. Determine a linear normalization factor for each spectrum that will minimize the summed difference between all intensities observed within the specified mass range from 710 Da to 27,990 Da and the corresponding intensities in the overall, baseline-corrected median spectrum.
   c. Because different spectra may have slightly different mass ranges, which would change the dimensions of the dataset, we normalize with respect to only those points in the mass range 710 to 27,990 Da, even if the actual data extend beyond this range.
   d. The normalization factor, f_n, is calculated by minimizing the difference between I_n and I_{median}, which requires finding a f_n to make $f_n \times I_n = I_{median}$. This involves a least-square regression analysis of all observed intensities in the specified mass range in the median curve vs those in each individual spectrum, with the slope of this curve = f_n.
   e. Re-normalize each spectrum's raw normalization value (as determined previously) by setting the median overall raw normalization value equal to 1.0, then delete from the dataset all spectra with re-normalized factors F below 0.5 or above 2.0 (the latter spectra should be flagged, individually examined, and if needed, be repeated, as they are likely to include poor spectra).

   f. If any spectra are eliminated via the above test, then re-normalize the remaining spectra.
   g. Multiply each background-corrected, data point intensity in each spectrum by this re-normalization factor.
7. Remove background noise that otherwise would lead to instability in the classification program and limit peak consideration to one mass-vs-intensity data point for each observed peptide ion isotope. Hence, each data point must pass the following criteria in order to be considered as a possible marker:
   a. Minimum intensity level test: rank the normalized intensities for each spectrum and take only data points whose intensities are within the top 20%.
   b. Peak test: assume that only data points in completely or partially resolved peaks result from peptide ions and are likely to be meaningful. Data points in partially resolved peaks may represent the intensity sum of a useful marker superimposed on an unrelated, nonmarker peptide ion.
      i. At least three of four successive data point intensities before or after each candidate marker data point must show a progressive increase or decrease in background corrected, normalized peak intensity.
      ii. Extend each peak by an additional four data points before/after the last data point that passed the peak test. Hence at this point we include a band width around each peak.
      iii. At least 10% of the cases or controls need to pass the peak test for this peak to be considered as a useful marker peak.
8. Unique peptide origin test: because there is no benefit in having multiple markers that all originate from different isotopes of the same peptide ion, all potential markers must have *m/z* values that differ from each other by at least 3.1 Da. Hence, if multiple potential markers are found that differ from each other by less than 3.1 Da, only the best marker (as judged by its individual ability to discriminate all cases from control samples in the training set) is considered. It is possible (e.g., if there are incompletely resolved, unrelated peptide ions that overlap with the C12 isotope peak of a marker peptide ion) that the best isotopic representative of a marker ion is not the C12 isotope. Without carrying out the unique peptide ion origin test, the variable importance measure determined by the RF algorithm would be unstable. That is, markers arising from different isotopes of the same peptide should be surrogates for each other. If one marker (i.e., from the C12 isotope) from a peptide is selected in a tree construction, then randomly permuting the values of the corresponding C13 isotope peak from this same peptide would not be expected to significantly change our classification ability. Hence, both these markers would appear to be unimportant. Instead, for each peptide we select only one mass-vs-intensity data point from whichever isotope (assuming the marker arises from a reflectron spectrum; note that this will not be a problem with linear spectra where peptide/protein isotopes are not resolved), which can best discriminate all cases from controls in the training set based on some criterion. For this purpose, we carry out an initial rough filtering, using the normalized difference between case/control group for all candidate markers within the 3.1-Da mass range to identify the best marker.

3.2. Feature Selection and Sample Classification

The data points obtained from the data preprocessing step then represent potential biomarkers. The following is a step-by-step outline of the basic features of a random forest algorithm *(1)* we have implemented to find markers that can best discriminate cases from control samples (*see* **Notes 8–11** for a review of different classification methods and our motivation for using the random forest algorithm):

1. Prespecify the number of trees to grow (currently, we are using 5000 trees), which will constitute our forest. Each tree will be grown as follows:

   Randomly divide the training sample into training/testing set-approx 67% as the training set, 33% as the testing set. Our tree will be constructed based on the training set according to the following procedure. Keep in mind that each training set is a resample with replacement from the original training sample; on average it contains approx 67% of the original training samples, and the remaining "testing" set contains approx 33% of the spectra in the training set.

   a. Start from all samples as one root node. Randomly select m_{try} variables without replacement from all variables (say 60 mz vs intensity values that passed the peak test and other tests described above); for each variable, find the best cutoff intensity value to best split the root node, which means to find the intensity cutoff value that will divide the maximum number of cases vs control samples correctly. This will then provide a measure of the decrease of impurity (i.e., two child nodes should contain a more homogeneous mix than the root node) from root node to two child nodes for each selected variable. Among these m_{try} variables, select the best one, based on the decrease of impurity, as our final split variable for the root node. As for measure of impurity, we use the Gini index, $1\text{-}(P_1\text{^}2 + P_2\text{^}2)$, where P_1 is the proportion of case samples, P_2 is the proportion of normal samples, and $P_1 + P_2 = 1$. When $P_1 = 0$ or 1, Gini is 0, which is least impure. When $P_1 = P_2 = 0.5$, Gini is 0.5, which corresponds to a maximum impure sample. Our overall goal is to reduce this Gini index, and the overall decrease of Gini index due to one variable can reflect the importance of this variable. After we select a split variable, we can split the root node into two child nodes, and the decrease of impurity is measured as: $P_1\text{^}2 + P_2\text{^}2 - (R_1\text{^}2 + R_2\text{^}2) - (L_1\text{^}2 + L_2\text{^}2)$, where P_1 and P_2 are the proportions of cancer/control for all samples in the root node, R_1/R_2 and L_1/L_2 are the proportions in the right/left child nodes, $P_1 + P_2 = R_1 + R_2 = L_1 + L_2 = 1$. This function is always positive, and we want to maximize it by choosing different cutoff values for selected split variables.
   b. For each child node, repeat the previous growing step as for the root node. We keep on growing until the tree perfectly fits the data, i.e., all terminal nodes contain only one class of samples.
   c. Repeat the above two steps 5000 times to get 5000 trees, with each tree having an associated training and testing set.
2. Use these 5000 trees to get a prediction for each sample and a measure of importance for each variable, as described in the following:

a. For each sample S, find those trees which are based on a training set not containing this sample, label this set as Pred, and suppose Pred has m trees. Put S down Pred to get m predictions for S. The final prediction for S will be the majority vote of these I>m predictions. For these m predictions, we will get a margin for S: margin = proportion of correct prediction - proportion of incorrect prediction (this assumes there are only two classes of samples; for >2 classes, we will minus the maximum of incorrect class predictions). Ideally we want margin to be as large as possible, and one of our importance measures will be based on this margin.
b. The importance measure for the mth variable is calculated as follows: For each tree, we have a testing set. Randomly permute the value of the mth variable in the testing set, and we can get a new prediction for testing samples. Proceed as previously for all 5000 trees. Finally, we can get a new prediction error and also a new margin for each sample. The decrease of margin after permutation will reflect the importance of the mth variable. The increase of prediction error can also serve as an importance measure, but intuitively this measure will be very crude, because essentially it loses some information by summarizing the continuous margin with a discrete value (0/1).
c. Select mk (e.g., 28) biomarkers based on the importance measure, and construct a random forest for these mk variables as previously. As a new sample comes in, use this forest to predict the class of this sample. Our confidence in the prediction can be measured with the margin.

3. Export a list of potential biomarkers. For each marker, plot the following:
   a. *m/z* vs intensity for all background corrected, normalized case vs all control samples for the *m/z* region extending for 3 Da on either side of the marker.
   b. Make a similar plot as described previously for every fifth case vs control sample.
   c. Overall median case vs control intensities around each marker, extending for 3 Da on either side of the marker.
4. Delete all potential markers that cannot reasonably contribute to a partially or completely resolved peptide isotopic cluster.
5. Use the resulting markers to develop, as discussed later, the quantitative criteria for the diagnosis of unknown samples, with the first step being normalizing the unknown sample with respect to the training set.
   a. For these selected markers, construct binary classification trees for future prediction. We grow 5000 trees instead of one tree. Basically, the more trees we have, the more stable the final prediction error is. We can run several times with 1000 to 10,000 trees to evaluate the effects. Typically, 5000 trees will give a very stable result. We use random selections in the tree construction process. Consequently, we can have different results, although the markers used are exactly the same for different trees.
   b. The following is one typical run of our construction process:
      i. Sample with replacement from all samples to form a new training set.
      ii. For this new training set, construct one tree. At each node split, we randomly select one *m/z* marker value to split.

iii. Repeat **steps i** and **ii** until desired number of nodes and trees is reached.
iv. For future samples, we do a majority vote using all of the trees.

3.3. Sample Split Analysis: How Will the Sample Size Affect Prediction Accuracy?

This can be addressed by estimating a "sample size" vs "prediction accuracy" curve. This basically requires the following procedure:

1. Suppose we have a very large population S, which contains all potential samples.
2. Given a sample size N.
3. Randomly take N samples from S to form a training set T_N, and construct a classifier $C(T_N, .)$ using our procedure.
4. Use $C(T_N, .)$ to classify a random sample from S, and calculate its error rate

$$E_{a \in S} C(T_N, a)$$

Here the expectation is taken over all potential samples from S.

5. Take the average over the previous error to get the error estimation for sample size N:

$$E_{|T_N| = N} E_{a \in S} C(T_N, a)$$

Generally, we do not have the full population S. Instead, we only have a finite sample, S_n (n = 187 for our ovarian cancer dataset described later), which will be used as an estimator of S. In practice, we use the following estimation procedure, based on a finite sample S_n:

a. Given sample size N (<n), we take N = 40, 80, 120 to get a rough estimate for the general trend of error vs sample size curve.
b. Randomly take N samples (without replacement) from S_n to form a training set T_N. Then construct a classifier $C(T_N, .)$.
c. Use $C(T_N, .)$ to classify S_n and get an error estimate

$$E = \frac{\sum_{i=1}^{n} C(T_N, a_i)}{n} = \frac{N}{n} E_0 + \frac{n-N}{n} E_1$$

where

$$E_0 = \frac{\sum_{a_j \in T_N} C(T_N, a_j)}{N}$$

is the total error rate we reported using our RF program, and

$$E_1 = \frac{\sum_{a_j \notin T_N} C(T_N, a_j)}{n - N}$$

is similar to an out-of-bag error estimation (error for samples not in training set).

d. Repeat **steps b** and **c** M times, e.g., M = 100, to get M error estimations E^1, ... , E^M
e. Use $(E^1 + ... + E^M)/M$ to estimate the final error rate for sample N.

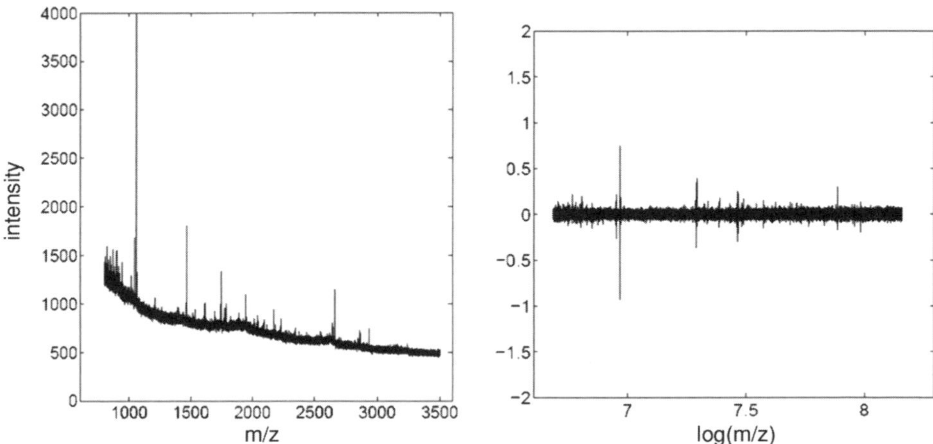

Fig. 3. Data preprocessing example. Left: original raw data. Right: mass spectrometry data after baseline correction.

3.4. Application Example

In this section, we apply the methods described in the previous section to an ovarian cancer dataset to identify cancer-related biomarkers. The MALDI-MS spectra were acquired as described previously on sera (obtained from the National Ovarian Cancer Early Detection Program at Northwestern University Hospital) from 77 control cases and 93 disease cases, using a Micromass M@LDI-L/R instrument, with 700–3500 Da for reflectron and 3450–28,000 Da for linear, and this resulted in a merged dataset consisting of MS spectra that extend from 700 to 28,000 Da.

Figure 3 shows how our workflow preprocesses the input raw data. On the left is an example of a MALDI-MS raw spectrum obtained on serum, the decreasing tendency is obvious. After taking a logarithm transformation on both the *m/z* and intensity axis, we estimate the baseline of the curve and subtract the baseline from the data. The resulting curve on the right plot clearly shows that the decreasing tendency has been removed.

After data preprocessing, we apply the random-forest method on the peaks to identify biomarkers that are most informative in terms of distinguishing disease cases from control cases. **Figure 4** gives a two-dimensional visualization of the samples based on the proximity matrix generated when we run the random-forest algorithm. In the graph, most samples are well separated, but some of them mix together.

Figure 5 shows the impact of training size on the classification rate, as well as the additional information offered by the data from the linear range. We can clearly see the trend that larger training sets result in smaller classification

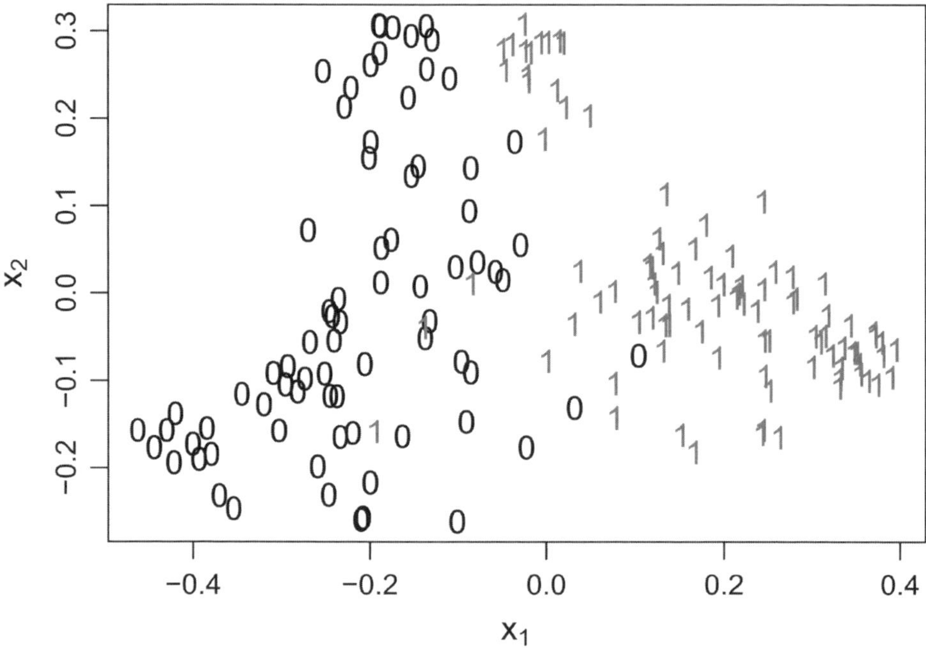

Fig. 4. The symbol "0" represents the samples from control group and "1" represents cancer group.

errors. And for a fixed training set, classification error drops dramatically from 5 to 20 biomarkers and then it levels off at approx 20–40 biomarkers for the combined reflectron + linear data. With 136 samples in the training set, we can achieve approx 20% classification error.

4. Notes

1. During peak alignment, simply using a linear shift may not work all the time, as previous studies have shown that the variation of peak locations in different datasets is nonlinear ***(3,4)***, and this variation still exists even for technical replicates (data not shown). The reasons that underlie data variation are extremely complicated, including differences in sample preparation, chemical noise, co-crystallization and deposition of the matrix-sample onto the MALDI-MS target, and laser position on the target. Here we are interested in reducing the variation and aligning these peaks together. As for peak alignment, we are exploring a superset idea that is very similar to the clustering approach in ***(5)***. Based on the assumption that the peak variation range is smaller than the minimal distance dmin, we construct a superset of all peaks as follows:
   a. Set a distance threshold as the smallest distance between neighboring peaks in all peak sets.
   b. Initialize the superset as the first peak set.

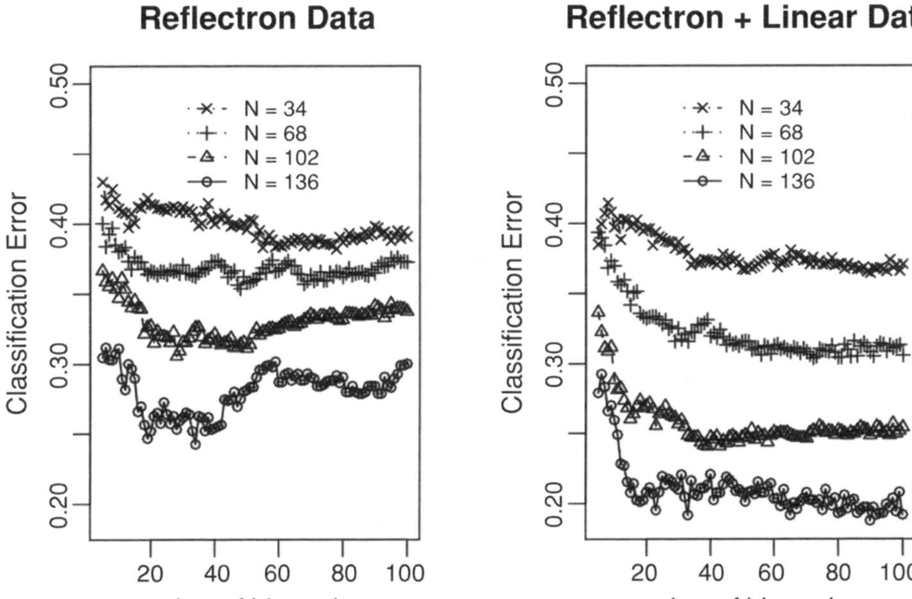

Fig. 5. The fivefold cross validation classification error estimations for our ovarian cancer data. After adding the linear data, the best classification error achieved drops from approx 25% to 20% and the classification error estimation is also more stable.

c. For each point in a new peak set, check its counterpart in the superset that has the smallest distance:
   i. Over the threshold: new peak, add to the super set;
   ii. Below the threshold: already in the superset, ignore.

Repeat **step 1c** until all new peak sets are processed. After the superset is constructed, we can align every peak in each peak set to the superset. It is guaranteed that each peak point in a peak set has at least one possible counterpart in the superset.

2. There are related approaches to address the peak alignment problem. Coombes et al. ***(6)*** pooled the list of detected peaks that differed in location by three clock ticks or by 0.05% of the mass. Yasui et al. ***(7)*** believed that the *m/z* axis shift of peaks are approx ±0.1% to ±0.2% of the *m/z* value. Thus, they expanded each peak to its local neighborhood, with the width equal to 0.4% of the *m/z* value of the middle point. This method certainly oversimplifies the problem. In another study ***(8)***, they first calculated the number of peaks in all samples, allowing certain shifts, and selected *m/z* values with the largest number of peaks. This set of peaks is removed from all spectra, and the procedure is iterated until all peaks are exhausted from all the samples. In a similar spirit, Tibshirani et al. ***(5)*** proposed to use complete linkage hierarchical clustering in one dimension to cluster peaks, and the resulting

dendrogram is cut off at a given height. All the peaks in the same cluster are considered as the same peak in further analysis. Randolph and Yasui ***(9)*** used wavelets to represent the MS data in a multi-scale framework. They used a coarse-to-fine method to first align peaks at a dominant scale and then refine the alignment of other peaks at a finer scale. From the signal representation point of view, this approach is very interesting. But it remains to be determined whether the multi-scale representation is biologically reasonable. Johnson et al. ***(10)*** assumed that the peak variation is less than the typical distance between peaks, and they used a closest point matching method for peak alignment. The same idea was also used in Yu et al. ***(11)*** to address the alignment of multiple peak sets. Certainly, this method is limited by the data quality and it cannot handle large peak variation. Dynamic programming (DP) based approaches ***(3,12)*** have also been proposed. DP has been used in gene expression analysis to warp one gene expression time series to another similar series obtained from a different biological replicate ***(13)***, where the correspondence between the two gene expression time series is guaranteed. In MS data analysis, however, the situation is more complicated, because a one-to-one correspondence between two datasets does not always exist. Although it is still possible to apply DP to deal with the short-of-correspondence problem, some modifications are necessary (such as adding an additional distance penalty term in the estimation of correspondence matrix). It also remains unclear how DP can identify and ignore outliers during the matching. Eilers ***(4)*** proposed a parametric warping model with polynomial functions or spline functions to align chromatograms. In order to fix warping parameters, he added calibration example sequences into chromatograms. Although the idea of using a parametric model is interesting, it is difficult to repeat the same parameter estimation method in MS data, because we cannot add many calibrator compounds into the MS samples. Also, it is unclear whether a second-order polynomial would be enough to describe the nonlinear shift of MS peaks.

3. Peak detection is an essential step in data preprocessing. The results in this chapter are obtained using a simple peak identification method. Right now, we are applying the amplitude modulation (AM) technique to form a peak envelope signal for each peptide. The AM technique is widely used in telecommunication and broadcasting, where a carrier with a fixed frequency f0 moves the original signal up to a transmitting frequency range. After receiving the transmitted signal, we use an envelope detection technique to reconstruct the original signal. Our idea here is to apply the envelope detection on the MS data before searching for the local maxima. In this way, the multiple peaks corresponding to the same peptide can be considered as a sampled version of a much smoother envelope signal. Envelope detection makes it much easier to group these successive peaks together, thus to reduce the complexity of the spectrum such that each peptide will be represented by only a single peak. It should be noted that applying a simple low-pass filter (like a Gaussian filter) will not guarantee the extraction of the envelope signal, because the carrier frequency f0 may move the envelope signal beyond the low-pass region. Consequently, the low-pass filtering will over-blur the signal. Here we apply a Gabor filter ***(14)*** centered at f0 on the data and take the amplitude of the filter response as output. The Gabor fil-

ter consists of an even real part (i.e., Gaussian weighted cosine function) and an odd imaginary part (i.e., Gaussian weighted sine function). Because the amplitude response is not sensitive to the phase variation, this process is also called quadrature filtering. We can then search for local maxima with a discrete differentiation method, followed by a peak width check, in which we keep only those local maxima whose left and right neighboring local minima have distances larger than a threshold value. In this algorithm, several parameters need to be determined, and they are different based on specific applications.

4. As we have mentioned, most ions produced in MALDI-MS have only single charges. Thus, the isotopically resolved peaks are assumed to have a constant distance of 1 Da. The equal spacing property of the MALDI data has been used in several approaches to identify peptides from isotopically resolved peaks. Breen et al. ***(15)*** used morphology and watershed algorithms to first detect a set of individual peaks and then use a Poisson distribution model (based on equal spacing between isotopic peaks) to detect peptides. They also demonstrated that their algorithm can separate two overlapping peptides using the Poisson isotopic distribution model. Similarly, Gras et al. ***(16)*** used a template-matching algorithm to identify peptides from a set of detected peaks, while they used the equal spacing property in the construction of their parameterized template. The same equal spacing property is the basis for extracting the envelope signal in this approach.
5. Envelope detection does not preserve the total intensity of a single peak, which is very important for quantitative studies. The point here, however, is to reduce the redundancy among multiple peaks that correspond to the same peptide. Guided by the envelope signal-based peak detection results, we still can go back to the data before envelope detection and detect different isotope peaks for the same peptide. Roughly speaking, this is a kind of top-down strategy, whereas previous approaches in ***(15,16)*** used a bottom-up strategy.
6. Normally, spectral peaks are local maxima in MS intensity data. In most published algorithms, peaks are mostly defined with respect to their nearby points. All these methods are based on similar intuitions and heuristics. Several parameters need to be specified beforehand in these algorithms, e.g., the number of neighboring points and the intensity threshold value. In fact, the parameter settings are related to our understanding/modeling of the underlying noise. Coombes et al. ***(6)*** defined noise as the median absolute value of intensity. Satten et al. ***(17)*** used the negative part of the normalized MS data to estimate the variance of noise. Wavelets-based approaches ***(18,19)*** have also been proposed to de-noise the MS data before peak detection. Based on the observation that there are substantial measurement errors in the intensity value, Yasui and his coworkers ***(7)*** argued that binary peak/nonpeak data is more useful than the absolute values of intensity, while they still used a local maximum search method to detect peaks. Clearly, the success of noise-estimation-plus-threshold methodology largely depends on the correctness of the noise model, which remains to be validated.
7. Another issue in peak detection is to avoid false-positive peaks. This is often done by adding an additional constraint (e.g., the peak width constraint ***[7]***) or by choosing a

specific scale level after wavelets decomposition of the original MS data (*(9)*). In the case of high-resolution spectra, it has been proposed that more than one isotopic variant of a peptide peak should be present before a spectral peak is considered to result from peptide ionization. It may also be possible to use prior information about the approximate expected peak intensity distribution of different isotopes arising from the same peptide during peak detection (i.e., the theoretical relative abundance of the first peptide isotope peak may range from 60.1% for polyGly (n = 23, MW 1329.5 Da) to 90.2% for poly Trp (n = 7, MW 1320.5 Da). Certainly, we also have to consider the issue of limited resolution and the consequent overlapping effect of neighboring peaks.

8. There are many classification methods, e.g., linear and quadratic discriminate analysis, and k–nearest neighbor. Most of these methods were developed in the pre-genome era, where the sample size was usually very large whereas the number of features was very small. One can intuit that the sample classification error rate will increase with more noise in the data. Therefore, directly applying these methods to proteomic datasets may not work. Instead, feature selection methods are usually applied to select some useful features at first, and then the selected features are used to carry out sample classification based on traditional classification methods. In this context, feature selection serves two purposes: for biological interpretation and for reducing the impact of noise. Two sample t-test statistics or variants thereof are often used to quantify the difference between two groups in the analysis of gene expression data (e.g., *see* **ref.** ***19***). However, determining the right dimension of features in feature selection still remains a difficult problem: if we only select a small number of features, we may miss many useful features. If we select a large number of features, we may over-fit the data with noise. Another problem with a t-statistic is robustness: the univariate feature selection based on a t-statistic is very sensitive to noise.
9. Instead of using univariate feature selection methods, it may be useful to consider multivariate feature selection methods. Lai et al. ***(20)*** analyzed the co-expression pattern of different genes to select genes that have differential gene-gene co-expression patterns with a target gene. Some interesting genes have been found to be significant and have been reported to be associated with prostate cancer, yet none of them showed marginal significant differential gene expressions. Generally, the multivariate feature selection is a combinatorial approach. To analyze two genes at a time, we need to consider n2 possibilities instead of n for the univariate feature selection. For analyzing the interaction of K genes we need to consider nK possibilities, this quickly becomes intractable.
10. Machine learning-based approaches have been proposed to couple feature selection and sample classification. They implicitly approach the feature selection problem from a multivariate perspective. The significance of a feature may highly depend on other features. Isabelle et al. ***(21)*** have reported using support vector machines (SVM) to select genes for cancer classification from microarray data. Qu et al. ***(22)*** applied a boosting tree algorithm to classify prostate cancer samples and to select important peptides using MS analysis of sera. Wu et al. ***(1)*** reported using random

forest to select important biomarkers from an ovarian cancer dataset based on MALDI-MS analysis of patient sera, and showed good performance of random-forest feature selection compared to univariate t-statistic selection.

11. For the random-forest algorithm from Breiman *(23)*, randomness is introduced at each node split. Specifically, at each node split, a fixed number of features is randomly selected from all the features, and the best split is chosen among these selected features. On the other hand, for the random subspace method developed by Ho *(24)*, a fixed number of features is selected at first and is used for the same original data to produce a tree classifier. Thus, both models have the effect of randomly using a fixed subset of features to produce a classifier, but differ in the underlying tree-building method.

Acknowledgments

This work was supported by National Heart, Lung and Blood Institute/National Institutes of Health (NIH) contract N01-HV-28186, National Institute on Drug Abuse/NIH grant 1 P30 DA018343-01, and NSF grant DMS-0241160.

References

1. Wu, B., Abbott, T., Fishman, D., et al. (2003) Comparison of statistical methods for classification of ovarian cancer using mass spectrometry data. *Bioinformatics* **19,** 1636–1643.
2. Petricoin, III, E., Ardekani, A. M., Hitt, B. A., et al. (2002) Use of proteomic patterns in serum to identify ovarian cancer. *Lancet* **359(9306),** 572–577.
3. Torgrip, R., Aberg, M., Karlberg, B., and Jacobsson, S. (2003) Peak alignment using reduced set mapping. *J. Chemomet.* **17,** 573–582.
4. Eilers, P. (2004) Parametric time warping. *Anal. Chem.* **76,** 404–411.
5. Tibshirani, R., Hastie, T., Narasimhan, B., et al. (2004) Sample classification from protein mass spectrometry, by "peak probability contrasts." *Bioinformatics* **20(17),** 3034–3044.
6. Coombes, K., Fritsche, Jr, H., Clarke, C., et al. (2003) Qualitycontrol and peak finding for proteomics data collected from nipple aspirate fluid by surface-enhanced laser desorption and ionization. *Clin. Chem.* **49,** 1615–1623.
7. Yasui, Y., Pepe, M., Thompson, M., et al. (2003) A data-analytic strategy for protein biomarker discovery: profiling of high-dimensional proteomic data for cancer detection. *Biostatistics* **4(3),** 449–463.
8. Yasui, Y., McLerran, D., Adam, B., Winget, M., Thornquist, M., and Feng, Z. (2003) An automated peak identification/calibration procedure for high-dimensional protein measures from mass spectrometers. *J. Biomed. Biotechnol.* **4,** 242–248.
9. Randolph, T. and Yasui, Y. (2004) Multiscale processing of mass spectrometry data, in *University of Washington Biostatistics Working Paper Series*, Number 230.
10. Johnson, K., Wright, B., Jarman, K., and Synovec, R. (2003) High-speed peak matching algorithm for retention time alignment of gas chromatographic data for chemometric analysis. *J. Chromatog. A* **996,** 141–155.

11. Yu, W., Wu, B., Lin, N., Stone, K., Williams, K., and Zhao, H. (2005) Detecting and aligning peaks in mass spectrometry data with applications to MALDI. *Comp. Biol. Chem.*, in press.
12. Nielsen, N., Carstensen, J., and Smedsgaard, J. (1998) Aligning of single and multiple wavelength chromatographic profiles for chemometric data analysis using correlation optimised warping. *J. Chromatog. A* **805,** 17–35.
13. Aach, J. and Church, G. (2001) Aligning gene expression time series with time warping algorithms. *Bioinformatics* **17,** 495–508.
14. Granlund, G. H. and Knutsson, H. (1995) *Signal Processing for Computer Vision.* Kluwer Academic Publishers.
15. Breen, E., Hopwood, F., Williams, K., and Wilkins, M. (2000) Automatic Poisson peak harvesting for high throughput protein identification. *Electrophoresis* **21,** 2243–2251.
16. Gras, R., Mueller, M., Gasteiger, E., et al. (1999) Improving protein identification from peptide mass fingerprinting through a parameterized multi-level scoring algorithm and an optimized peak detection. *Electrophoresis* **20,** 3535–3550.
17. Satten, G., Datta, S., Moura, H., et al. (2004) Standardization and denoising algorithms for mass spectra to classify whole-organism bacterial specimens. *Bioinformatics* **20(17),** 3128–3136.
18. Coombes, K., Tsavachidis, S., Morris, J., Baggerly, K., Hung, M., and Kuerer, H. (2004) Improved peak detection and quantification of mass spectrometry data acquired from surface-enhanced laser desorption and ionization by denoising spectra with the undecimated discrete wavelet transform. Tech. rep., The University of Texas M.D. Anderson Cancer Center.
19. Dudoit, S., Yang, Y. H., Speed, T. P., and Callow, M. J. (2002) Statistical methods for identifying differentially expressed genes in replicated cdna microarray experiments. *Statistica Sinica* **12,** 111–139.
20. Lai, Y., Wu, B., Chen, L., and Zhao, H. (2004) Statistical method for identifying differential gene-gene coexpression patterns. *Bioinformatics* **20,** 3146–3155.
21. Isabelle, G., Jason, W., Stephen, B., and Vladimir, V. (2002) Gene selection for cancer classification using support vector machines. *Mach. Learning* **46(1–3),** 389–422.
22. Qu, Y., Adam, B.-L., Yasui, Y., Ward, M. D., Cazares, L. H., Schellhammer, P. F., et al. (2002) Boosted decision tree analysis of surface-enhanced laser desorption/ionization mass spectral serum profiles discriminates prostate cancer from noncancer patients. *Clin. Chem.* **48(10),** 1835–1843.
23. Breiman, L. (2001) Random forests. *Mac. Learning* **45,** 5–32.
24. Ho, T. (1998) The random subspace method for constructing decision forests. *IEEE Trans. Pattern Anal. Mach. Intell.* **20,** 832–844.

15

Using the Global Proteome Machine for Protein Identification

Ronald C. Beavis

Summary

This chapter describes the use of an open-source, freely available informatics system for the identification of proteins using tandem mass spectra of peptides derived from an enzymatic digest of a mixture of mature proteins. The chapter describes the use of features of the Global Proteome Machine (GPM) interface that assist in making comprehensive assignments between spectra and sequences, including the detection of point mutations, posttranslational modifications, and experimental artifacts. The use of this interface to validate results using the GPM Database is also described. This data repository allows analysts to compare their own results to those obtained by other scientists to determine the degree to which their data are consistent with previous measurements.

Key Words: Bioinformatics; protein identification; peptide sequencing; spectrum interpretation.

1. Introduction

Proteomics has become dominated by the ability to associate the proteins in a sample with previously observed protein sequences. The older, more established immunological techniques for performing this type of assignments, such as Western blotting or enzyme-linked immunosorbent assay (ELISA), have become secondary validation techniques. The faster, easily automated, lower unit cost mass spectrometry (MS)-based methods currently dominate the research in the literature. Of the MS-based techniques, the use of trypsin digestion followed by reverse-phase chromatography with an online tandem mass spectrometer as a detector has become by far the most widely used and accepted instrumental protocol *(1–3)*.

The bioinformatics approaches being used for performing the data analysis can be divided into two approaches: those that require interpretation of the mass list (Class A) *(4)*, frequently referred to as *sequence mass tagging*, and those

From: *Methods in Molecular Biology, Vol. 328: New and Emerging Proteomic Techniques*
Edited by: D. Nedelkov and R. W. Nelson © Humana Press Inc., Totowa, NJ

that require no interpretation (Class B) *(5)*. Both types of algorithms have been used since very early in the development of algorithms for MS-base proteomics. They have also both been used to develop commercial software products.

Class B type analysis has become a commonplace method of determining the sequence of a peptide. This analysis is done by comparing a spectrum with all of the peptide sequences that are known for a particular organism's proteome. Until recently, it was necessary to use proprietary algorithms and software to perform this analysis. The development of open-source informatics has revolutionized the field *(6,7)*.

The Global Proteome Machine (GPM) *(8)* is one of the new open-source systems. It utilizes a Class B open-source search engine, X! Tandem *(6)*. It uses a new type of user interface, which allows ready access to genomic information and literature resources. It also incorporates information obtained from a repository of experimental information that allows the immediate comparison of a particular result with those obtained in other laboratories. This repository allows investigators to benchmark their results against those of other groups, as well as to use the details of others' results to plan their own experiments.

The X! Tandem search engine uses a different strategy for performing searches *(9)*. Rather than attempting to compare a collection of tandem mass spectra against all possible variations of all proteins in a proteome, the search process is broken up into two rounds of calculation. In the "survey" round, the spectra are compared against all of the sequences in the proteome, but with a limited number of potential modifications and with the assumption that the proteolytic enzyme used to produce peptides generated an ideal set of peptides-no missed or unexpected bond cleavages. The "refinement" round takes the protein sequences that were identified as having statistically significant matches to the spectra and performs matches using only these sequences, using multiple sets of search parameters that explore many possible post-translational and artifact side-chain modifications, unanticipated peptide bond cleavages, and amino acid point mutations.

2. Materials

1. An Internet-connected computer (PC or Macintosh).
2. An HTTP browser (e.g., Internet Explorer, Firefox, Safari).
3. A Scalar Vector Graphics (SVG) plugin for your browser (e.g., Adobe, Corel).
4. A collection of tandem mass spectra, organized in a standard file format (e.g., mzXML, mzData, DTA, or Mascot Generic Format).

3. Methods

3.1. Assigning Mass Spectra to Protein Sequences

1. Enter the universal resource locator (URL) of a GPM server, e.g., *http://h.thegpm.org*. The main user data input page (**Fig. 1**) should appear in the browser window (*see* **Notes 1** and **2**).

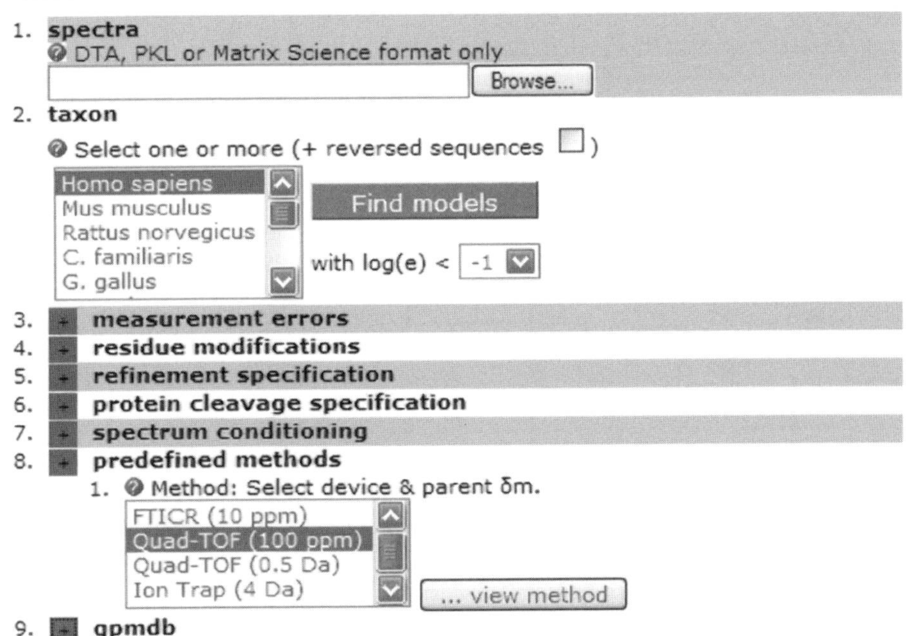

Fig. 1. The Global Proteome Machine experimental data query interface. Each section of the search form can be expanded (or hidden) by clicking on the appropriate "+" symbol next to the section header.

2. Select a set of default parameters associated with the type of tandem mass spectrometer that you used to collect the results, from the "Predefined methods" selection box (*see* **Note 3**).
3. Enter an appropriate value for the fragment ion mass tolerance. The quality of a spectrum-to-peptide match depends strongly on the fragment ion mass tolerance value being set correctly (*see* **Note 4**).
4. Locate your input spectrum data file on your computer, by using the "Browse" button (*see* **Note 5**).
5. Select the appropriate cleavage reagent that was used to create peptides from the intact protein samples. If you are using an enzyme or cleavage chemistry that is not listed, you will need to go to the "Advanced" search page and specify the cleavage chemistry. For example, trypsin's cleavage is specified by the string "[KR]|{P}", which is interpreted as cleavage at any peptide bond between either a lysine (K) or arginine (R) residue and any other residue, except proline (P). Square brackets indicate allowed residues and french brackets indicate forbidden residues. If there is no specificity for residues, then "X" is used, e.g., "[X]|[X]" is interpreted as cleavage at any peptide bond.

6. Select the appropriate complete side-chain modifications for your sample. Cysteine residues are often modified completely to improve peptide recovery: common cysteine modification chemistries are listed in a menu. If your chemistry is not present, you may specify it. These modifications will be used for both the survey and refinement rounds of analysis. Modifications that are not shown on the interface can be added manually, using the following format: "*m@r*", where *r* is the single letter abbreviation for the residue to be modified and *m* is the mass difference between the modified and normal residue. For example, oxidized methionine (the addition of a single oxygen atom, $m = 16$) would be indicated by "*16@M*".
7. Select the appropriate potential side-chain modifications for the survey round. These modifications can be biologically relevant, such as phosphorylation or acetylation, or they may be experimental artifacts, such as methionine oxidation or asparagine deamidation. It is recommended that you specify only very common modifications in the survey round, as the addition of many types of modification at this step can be very time consuming. These modifications are tested for every occurrence of the specified residue type.
8. Specify the potential modifications desired for the refinement rounds of identification. Sets of predefined modifications are available through a selection box. Multiple modifications can be set by holding down the "Alt" key and clicking the desired entries. Other modifications can be set manually. These modifications are tested for every occurrence of the specified residue type.
9. Specify potential modification motifs for the refinement rounds of identification. These motifs allow potential modifications at sequence-specific residues. For example, the motif specification "*80@[ST!]XX[DE]*" allows for a modification of 80 Daltons (characteristic of phosphorylation) at S or T residues, only if they are followed by two residues of any type followed by a D or E residue. The residue to be modified is indicated by an exclamation point.
10. Choose whether or not the proteins selected for refinement should be tested for unanticipated peptide-bond cleavage. These cleavages can be caused by endo- or exopeptidases in the original protein extract or contaminating enzymatic activity in the selected cleavage reagent, e.g., chymotryptic activity in a trypsin preparation.
11. Choose whether or not the proteins selected for refinement should be tested for single amino acid point mutations. If selected, each peptide will be checked against the input spectra with every possible single amino acid substitution for that peptide sequence.
12. Choose whether or not the spectra should be filtered for redundancy, prior to searching ("spectrum conditioning" section). If this option is selected, all of the spectra in a set will be compared, and spectra that appear to be repeats will be flagged and removed prior to searching. The most intense example of any set of repeated spectra will be retained.
13. Select the species or list of species that you would like to use for your search from the "Taxon" selection box. This species should correspond to the organism used to obtain the protein samples analyzed. Multiple species can be selected by holding down the "Alt" key and clicking on multiple entries in this box (*see* **Note 6**).

rank	log(e)	log(I)	%	#	total	M_r	accession
1	-13.6	3.10	8.5	3	4	34.3	ENSP00000339027 gpmDB homologues protein 60S ACIDIC RIBOSOMAL PROTEIN P0 (L10E). [Source: SWISSPROT (P05388)] Annotated domains: IPR001813 60S Acidic ribosomal protein IPR001790 Ribosomal protein L10
2	-9.3	3.67	13	3	4	24.2	ENSP00000307889 gpmDB homologues protein 60S RIBOSOMAL PROTEIN L13 (BREAST BASIC CONSERVED PROTEIN 1). [Source: SWISSPROT (P26373)] Annotated domains: IPR001380 Ribosomal protein L13e IPR001472 Bipartite nuclear localization signal
3	-7.9	3.67	20	2	4	14.0	ENSP00000332194 gpmDB homologues protein HISTONE H2A.Q (H2A/Q) (H2A-GL101) (FRAGMENT). [Source: SWISSPROT (Q16777)] Annotated domains: IPR007124 Histone-fold/TFIID-TAF/NF-Y IPR002119 Histone H2A IPR007125 Histone core
4	-7.0	3.10	4.2	2	2	47.5	ENSP00000346035 gpmDB homologues protein 60S ribosomal protein L4 (L1). [Source: SWISSPROT (P36578)] Annotated domains: IPR002136 Ribosomal protein L4/L1e IPR001472 Bipartite nuclear localization signal IPR002086 Aldehyde dehydrogenase
5	-6.7	3.07	8.1	2	3	29.3	ENSP00000260536 gpmDB homologues protein 60S RIBOSOMAL PROTEIN L7. [Source: SWISSPROT (P18124)] Annotated domains: IPR001472 Bipartite nuclear localization signal IPR000517 Ribosomal protein L30

Fig. 2. The results page describing the proteins found by a single search. Each of the proteins found in the search are displayed in a separate row, in order of the confidence of assignment.

14. Click on the "Find models" button. The GPM refers to the interpreted results from a collection of spectra as a "model" of the data.

3.2. Interpreting the Results of an Assignment

1. Once a search has been completed, the "model" page will be displayed automatically (**Fig. 2**). This page shows the results of a search, with a series of clickable links that allow the data to be displayed in a number of different views. The model page shows the overall results, tabulated so that the most confident protein assignments are listed at the top of the page. Confidence is expressed in terms of expectation values: an expectation value of 0.01 implies that there would be 1 assignment of this quality every 100 times if the assignment were random (*see* **Note 7**).
2. The headings for the model page are as follows:
   a. Rank: an ordinal number, in decreasing order of assignment confidence.
   b. log(e): the base-10 logarithm of the expectation value for the protein assignment.
   c. log(I): the base-10 logarithm of the summed intensity of all of the fragment ion mass spectra associated with this assignment.
   d. #: the number of peptides unique to the protein sequence assigned to spectra.
   e. Total: the number of spectra assigned to the protein.
   f. M_r: the molecular mass of the intact protein sequence, calculated assuming all ^{12}C isotopes.
   g. Accession: the appropriate accession number for the assigned protein (*see* **Note 6**).

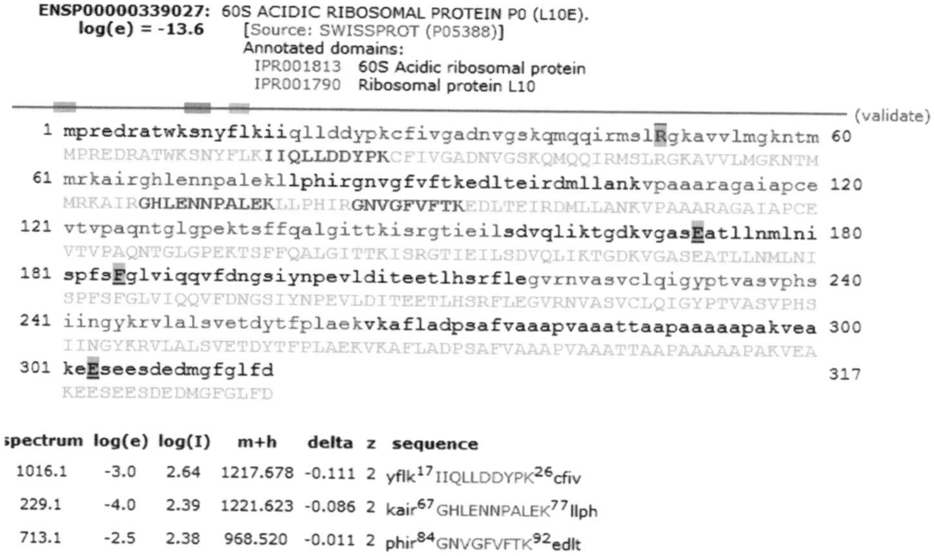

```
ENSP00000339027:  60S ACIDIC RIBOSOMAL PROTEIN P0 (L10E).
    log(e) = -13.6    [Source: SWISSPROT (P05388)]
                      Annotated domains:
                        IPR001813  60S Acidic ribosomal protein
                        IPR001790  Ribosomal protein L10
                                                                  (validate)
  1 mpredratwksnyflkiiqllddypkcfivgadnvgskqmqqirmslRgkavvlmgkntm  60
    MPREDRATWKSNYFLKIIQLLDDYPKCFIVGADNVGSKQMQQIRMSLRGKAVVLMGKNTM
 61 mrkairghlennpaleklIphirgnvgfvftkedlteirdmllankvpaaaragaiapce 120
    MRKAIRGHLENNPALEKLLPHIRGNVGFVFTKEDLTEIRDMLLANKVPAAARAGAIAPCE
121 vtvpaqntglgpektsffqalgittkisrgtieilsdvqliktgdkvgasEatllnmlni 180
    VTVPAQNTGLGPEKTSFFQALGITTKISRGTIEILSDVQLIKTGDKVGASEATLLNMLNI
181 spfsFglviqqvfdngsiynpevlditeetlhsrflegvrnvasvclqigyptvasvphs 240
    SPFSFGLVIQQVFDNGSIYNPEVLDITEETLHSRFLEGVRNVASVCLQIGYPTVASVPHS
241 iingykrvlalsvetdytfplaekvkafladpsafvaaapvaaattaapaaaaapakvea 300
    IINGYKRVLALSVETDYTFPLAEKVKAFLADPSAFVAAAPVAAATTAAPAAAAAPAKVEA
301 keEseesdedmgfglfd                                            317
    KEESEESDEDMGFGLFD
```

spectrum	log(e)	log(I)	m+h	delta	z	sequence
1016.1	-3.0	2.64	1217.678	-0.111	2	yflk^{17}IIQLLDDYPK^{26}cfiv
229.1	-4.0	2.39	1221.623	-0.086	2	kair^{67}GHLENNPALEK^{77}llph
713.1	-2.5	2.38	968.520	-0.011	2	phir^{84}GNVGFVFTK^{92}edlt

Fig. 3. Individual result for a single protein. The protein sequence, available exon-intron boundary, and single nucleotide polymorphism information, as well as the detailed assignment of peptides found during the search, are displayed.

3. The accession heading has several clickable links, which lead to specific displays:
   a. gpmdb: lead to a display that shows all occurrences of this particular accession number in the GPMDB database.
   b. Homologs: displays sequences that were assigned, but which appear to be homologous (*see* **item 4** below).
   c. Protein: displays the details of the protein's assignment (*see* **item 5** below).
4. The homolog display is composed by inspecting the list of proteins assigned, from best to worst expectation value. Starting at the best assignment, all other assignments that use the same spectra as the best assignments are declared as homologs of the best assignment. If these homologs do not have any other spectra assigned to them, then their accession numbers occur only on the homolog list. If they do have additional spectra, they appear on the model list, with an (H) prefix added to the accession number. It should be noted that this interpretation of *homolog* may have somewhat different results than a definition based solely on the protein's sequence.
5. The "protein" display shows the details of the assignment of a set of mass spectra to a protein sequence (**Fig. 3**). The display is composed of five sections:
   a. A tool bar giving clickable links to additional information about the protein, if available.
   b. A protein coverage map (*see* **item 8**).
   c. A display of the protein sequence (*see* **item 6**).
   d. A display of the peptides assigned to the sequence (*see* **item 7**).
   e. A description of the biological context of that protein, if available.

6. The protein sequence section of the display can be in two different forms. The simplest form shows the sequence with the assigned portions of the sequence in dark red and the unassigned portions in light red. The dark red sequence can be clicked, to directly show the peptide assignment details. This simple form of display is used when detailed gene model information is not available for a sequence or for organisms where the exon-intron structure is either very simple (e.g., brewer's yeast), or nonexistent (e.g., all prokaryotes). The second form shows the sequence on two lines, with the known genomic sequence above and the assigned sequence as follows. The genomic sequence indicates the exon structure of the gene model by alternating between black and blue letters, with exon boundaries that involve both exons in the nucleic acid triplet coding for residues in red. Known single nucleotide polymorphisms are indicated with red-background cells framing the appropriate residue. Placing the mouse cursor over these cells will produce a display showing the possible amino acid substitutions caused by the polymorphism. Known polymorphisms that do not produce amino acid substitutions are indicated with a green background. If a point mutation has been detected during the search, the position of the point mutation is indicated with a red-background cell on the assigned protein sequence (*see* **Note 8**).
7. The assigned peptide portion of the display sets out the peptide sequences that have been assigned to spectra in a table. By default, only the best assignments to a particular peptide sequence are listed; a full listing can be obtained by clicking on the "show all" link at the top of the page. The peptide sequences are ordered from the N-terminus to the C-terminus of the protein. The tabular headings are as follows:
   a. Spectrum: a number corresponding to the position on the original spectrum list of the spectrum assigned to this sequence. (Note: if a spectrum is assigned to more than one peptide sequence, this is indicated by the number following the decimal place.)
   b. log(e): the expectation value for the spectrum-to-sequence assignment.
   c. log(I): the summed intensity of the spectrum.
   d. m+h-the protonated molecular mass of the sequence, assuming all ^{12}C isotope composition.
   e. Delta: the difference between the measured and calculated protonated molecular mass, in Daltons.
   f. : the charge of the parent ion used for fragmentation.
   g. Sequence: the peptide sequence, shown in the context of the full protein sequence (this sequence is a clickable link to the "peptide" display; *see* **item 9**).
8. An important feature of the protein display is the sequence coverage map, displayed just above the protein sequence. This map, a line with a set of variable-intensity red bars, indicates the portions of the protein sequence have been observed in this experiment. The intensity of a red bar indicates the confidence of assignment: the darker the bar, the more confident the assignment. For a variety of reasons, not all portions of a protein sequence may be amenable to proteomic analysis. Certain peptide sequences seem to have the appropriate physical properties to be very readily observed, and the pattern of which peptides are most often observed is a characteristic of a protein sequence. These readily observed peptides

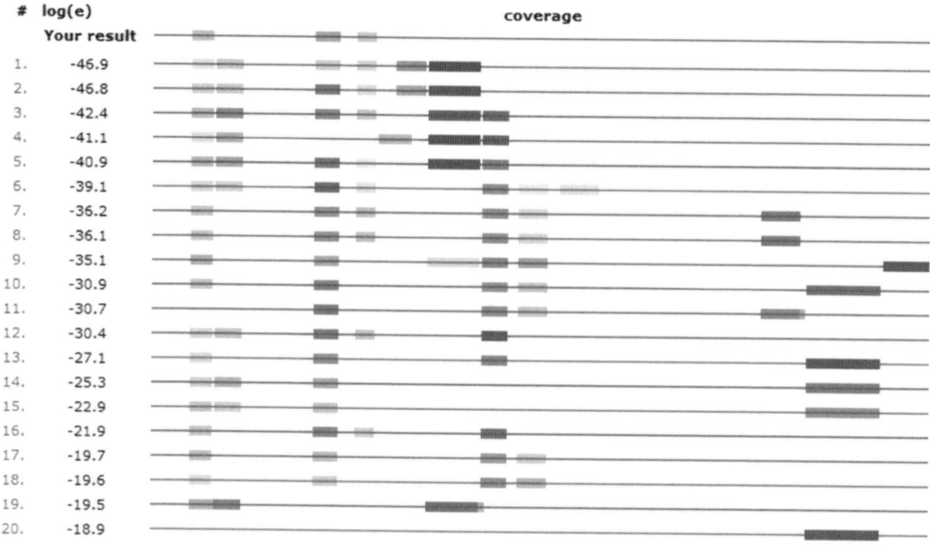

Fig. 4. Protein validation page derived from the protein result page (**Fig. 3**). The coverage map for the current result is displayed on the top of the diagram, followed by the coverage maps of the best assignments to the same protein sequence available in the Global Proteome Machine data repository.

are referred to as proteotypic peptides *(10)*. The "validate" link adjacent to the protein coverage map will produce a display from the GPMDB of the best assignments that have been made between spectra and the protein sequence in question (**Fig. 4**). The figure displays the coverage map of the current run on top, with maps from stored data displayed below, in descending order of confidence. Inspection of this diagram can be used to validate that an assignment conforms to previous data. This type of comparison is particularly useful when an assignment contains a limited number of peptide sequences.

9. The peptide sequence displayed shows the portion of sequence assigned in blue, with the sequence immediately N-terminal and C-terminal to the peptide shown in black. The residue numbering is relative to the N-terminus of the protein. Any residue that was chemically modified is represented by a blue-background cell, and any residue detected as a point mutation has a blue-background cell. Placing the mouse cursor over the colored cell will produce a display describing the modification(s) or mutation. Clicking on the blue peptide sequence will pop up a new window, showing the peptide display (*see* **item 10**).
10. The peptide display shows the details of one spectrum-to-sequence assignment (**Fig. 5**). The display has four sections:
    a. A toolbar with links to information about the peptide sequences (if available).
    b. A short description of the peptide sequence, describing modifications and mutation details (if applicable).

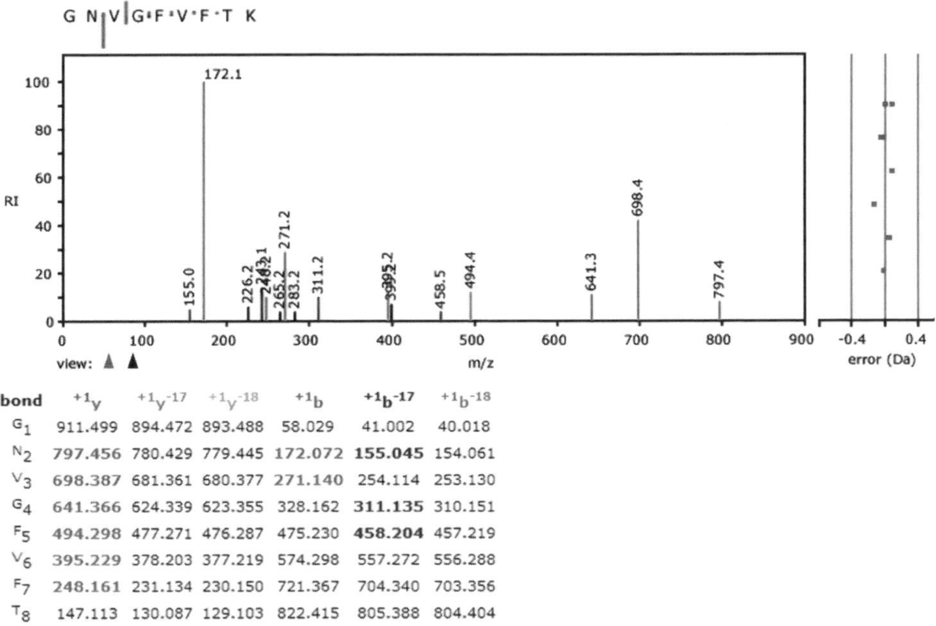

bond	+1y	+1y-17	+1y-18	+1b	+1b-17	+1b-18
G_1	911.499	894.472	893.488	58.029	41.002	40.018
N_2	**797.456**	780.429	779.445	**172.072**	**155.045**	154.061
V_3	**698.387**	681.361	680.377	**271.140**	254.114	253.130
G_4	**641.366**	624.339	623.355	328.162	**311.135**	310.151
F_5	**494.298**	477.271	476.287	475.230	**458.204**	457.219
V_6	**395.229**	378.203	377.219	574.298	557.272	556.288
F_7	**248.161**	231.134	230.150	721.367	704.340	703.356
T_8	147.113	130.087	129.103	822.415	805.388	804.404

Fig. 5. Detailed display of a spectrum-to-peptide sequence assignment. A fragmentation diagram, annotated spectrum, and fragment ion mass deviations are shown graphically, followed by a table showing the possible theoretical fragment masses for the sequence. The color coding is consistent between the table and the graphs.

c. A graphical display of the mass spectrum used to make the assignment (*see* **item 11** below).
d. A table of potential fragment masses, with the peaks assigned in the spectrum highlighted in bold text in color.

11. The graphical display of the mass spectrum has three component sections. The color scheme used for the table of fragment masses is used to indicate assignments in each of these component sections. These sections are as follows:
    a. A fragmentation diagram (top left-hand corner) showing the intensity of assigned peaks, plotted on the peptide sequence, with N-terminal fragments shown below the sequence and C-terminal fragments shown above.
    b. A conventional mass spectrum (center), with the peaks assigned to the peptide sequence marked in color (unassigned peaks shown in black).
    c. An error scatter plot (right-hand side) showing the difference between the observed and calculated masses for each assigned peak.
12. The "validate" link on this page leads to a selection of spectrum-to-sequence assignments from the GPMDB for this particular peptide sequence, with the graphical displays for each assignment stacked on top of each other (not shown). The default display has the 10 best assignments, arranged so that the spectrum most like

the current observed spectrum is displayed immediately below the current result. These assignments can be used to validate your result, as the pattern of peptide-bond cleavage is a property of the peptide sequence and is reproducible between different instruments. The spectra shown for validation all have the same charge state as the current spectrum; the fragmentation pattern for a sequence is dependent on the parent charge state (*see* **Note 9**).

4. Notes

1. The GPM interface and underlying informatics can be downloaded for free from http://www.thegpm.org.
2. The GPM interface is subject to change, from time to time. Various parameters discussed previously may be available on either the simple or advanced user interfaces.
3. The default settings for most parameters can be used for common types of analysis. Instructions and explanations for each of the parameters can be obtained by clicking the "?" icon associated with each entry box.
4. The performance of a mass spectrometer is often quoted as its mass accuracy in parts per million (ppm). A specification of ±20 ppm means that for an ion of *m/z* = 1000.0, the error associated with that mass will be 1000 × 0.000020 = ±0.020 Da. There is a temptation to over-specify the accuracy of parent ion mass measurements, based on this type of instrument specification, leading to significant missed identifications. These missed identifications have several causes. It is assumed that the all ^{12}C peaks (A_0) in the isotope cluster corresponding to a peptide have been correctly determined. Most software actually assigns the most abundant peak in the isotope cluster, which is often not the A_0 peak. This leads to systematic errors of 1 or 2 Da. Overconfidence can lead to misinterpretation, such as the assignment of nonexistent deamidations of asparagines or glutamine residues, which compensate for the systematic error in mass assignment. It is important to note that very accurate determination of a parent ion mass is not an important parameter for peptide identifications. The pattern of fragment ion masses is used to assure a good identification, not the parent ion mass. X! Tandem can compensate for this effect, by estimating which parent ions are likely to be improperly assigned and using that information to correct peptide sequence assignments.
5. When performing large searches, which require large amounts of calculation time, it is good practice to test a subset of the spectra first, to determine the appropriate parent ion and fragment ion mass tolerances that will generate the best results.
6. The proteome sequences available through the public GPM interface correspond to the protein translations of curated genomic or cDNA sequence collections. The use of sequence collections that are simply lists of all protein sequences that may have been reported for a species, such as the US National Center for Bioinformatics nonredundant (nr) collection, is strongly discouraged. These phenomenological lists contain many repetitions of some sequences (such as immunoglobulin chains) and they over-represent certain sequences that have been the subject of cross-

species evolutionary studies, such as animal oxygen carriers or plant photosynthesis-related proteins. More model species are added as new genomic resources become available.

7. The GPM attempts to use the latest information about a particular protein to produce its displays. Therefore, the system does not store a permanent record of the annotation associated with a particular protein accession number. When a new accession number is identified by the system, a query is made to the primary data source for that type of accession number (e.g., ENSEMBL ***(11)*** for human sequences or SGD ***(12)*** for yeast) and a local cached copy of this information is kept on that server. The protein display page for a particular result includes a button just above the external protein information section that allows the user to refresh the external information stored in the cache, to get the latest annotation for that protein accession.
8. Interpreting results that allow for point mutations in the peptide sequence can be difficult. If a sequence appears to have many point mutations, this is probably caused by unanticipated chemical modifications of residue side chains. Inspection of the mass shifts associated with the "point mutations" will often suggest the identity of the modification chemistry, e.g., unexpected amino group formylation (+27 Da) or acetylation (+42 Da). Point mutations do occur naturally, but there should be no more than one or two in a protein sequence, if the proteome examined is from the same species as the sample.
9. In addition to using the GPMDB data to validate new results, the database system has many other potential uses in proteomics. It can be search by keyword, peptide sequence, peptide mass, protein accession number, or GPM model number (a unique model number is assigned to each set of spectra searched). Information regarding the regions of a protein that can be observed using proteomics, likely homologs of a protein, proteins commonly observed along with a particular protein, and known potential side-chain modifications can all be determined, prior to performing an experiment. The GPM model accession numbers can be used for future reference, and links through to specific results in the GPMDB are encouraged.

References

1. Mann, M., Hendrickson, R. C., and Pandey, A. (2001) Analysis of proteins and proteomes by mass spectrometry. *Ann. Rev. Biochem.* **70,** 437–473.
2. Aebersold R. and Goodlett D. R. (2001) Mass spectrometry in proteomics. *Chem. Rev.* **101,** 269–295.
3. Wysocki, V. H., Resing, K. A., Zhang, Q., and Cheng, G. (2005) Mass spectrometry of peptides and proteins. *Methods* **35,** 211–222.
4. Mann, M. and Wilm, M. (1994) Error-tolerant identification of peptides in sequence databases by peptide sequence tag. *Anal. Chem.* **66,** 4390–4399.
5. Yates, J. R., 3rd, Eng, J. K., McCormack, A. L., and Schieltz, D. (1995) Method to correlate tandem mass spectra of modified peptides to amino acid sequences in the protein database. *Anal. Chem.* **67,** 1426–1436.

6. Craig, R., Cortens, J. P., and Beavis, R. C. (2004) An open source system for analyzing, validating and storing protein identification data. *J. Proteome Res.* **3,** 1234–1242.
7. Fenyö, D. and Beavis, R. C. (2003) A method for assessing the statistical significance of MS-based protein identifications using general scoring schemes. *Anal. Chem.* **75,** 768–774.
8. Craig, R. and Beavis, R. C. (2003) A method for reducing the time required to match protein sequences with tandem mass spectra. *Rapid Commun. Mass Spectrom.* **17,** 2310–2316.
9. Craig, R. and Beavis, R. C. (2004) TANDEM: matching proteins with tandem mass spectra. *Bioinformatics* **20,** 1466–1467.
10. Aebersold, R. (2003) Constellations in a cellular universe. *Nature* **422,** 115–116.
11. Birney, E., Andrews, T. D., Bevan, P., et al. (2004) An Overview of ENSEMBL. *Genome Res.* **14,** 925–928.
12. Issel-Tarver, L., Christie, K. R., Dolinski, K., et al. (2002) *Saccharomyces* genome database. *Meth. Enzymol.* **350,** 329–346.

Index

T

W–Z